Theory and Computation of Tensors

Theory and Computation of Tensors

Multi-Dimensional Arrays

WEIYANG DING
YIMIN WEI

AMSTERDAM • BOSTON • HEIDELBERG • LONDON
NEW YORK • OXFORD • PARIS • SAN DIEGO
SAN FRANCISCO • SINGAPORE • SYDNEY • TOKYO

Academic Press is an imprint of Elsevier

Academic Press is an imprint of Elsevier
125 London Wall, London EC2Y 5AS, UK
525 B Street, Suite 1800, San Diego, CA 92101-4495, USA
50 Hampshire Street, 5th Floor, Cambridge, MA 02139, USA
The Boulevard, Langford Lane, Kidlington, Oxford OX5 1GB, UK

Library of Congress Cataloging-in-Publication Data
A catalog record for this book is available from the Library of Congress

British Library Cataloguing-in-Publication Data
A catalogue record for this book is available from the British Library

ISBN 978-0-12-803953-3 (print)
ISBN 978-0-12-803980-9 (online)

For information on all Academic Press publications
visit our website at https://www.elsevier.com/

Working together
to grow libraries in
developing countries

www.elsevier.com • www.bookaid.org

Publisher: Glyn Jones
Acquisition Editor: Glyn Jones
Editorial Project Manager: Anna Valutkevich
Production Project Manager: Debasish Ghosh
Designer: Nikki Levy

Typeset by SPi Global, India

Preface

This book is devoted to the theory and computation of tensors, also called hyper-matrices. Our investigation includes theories on generalized tensor eigenvalue problems and two kinds of structured tensors, Hankel tensors and $\mathcal{M}$-tensors. Both theoretical analyses and computational aspects are discussed.

We begin with the generalized tensor eigenvalue problems, which are regarded as a unified framework of different kinds of tensor eigenvalue problems arising from applications. We focus on the perturbation theory and the error analysis of regular tensor pairs. Employing various techniques, we extend several classical results from matrices or matrix pairs to tensor pairs, such as the Gershgorin circle theorem, the Collatz-Wielandt formula, the Bauer-Fike theorem, the Rayleigh-Ritz theorem, backward error analysis, the componentwise distance of a nonsingular tensor to singularity, etc.

In the second part, we focus on Hankel tensors. We first propose a fast algorithm for Hankel tensor-vector products by introducing a special class of Hankel tensors that can be diagonalized by Fourier matrices, called anti-circulant tensors. Then we obtain a fast algorithm for Hankel tensor-vector products by embedding a Hankel tensor into a larger anti-circulant tensor. The computational complexity is reduced from $\mathcal{O}(n^m)$ to $\mathcal{O}(m^2 n \log mn)$. Next, we investigate the spectral inheritance properties of Hankel tensors by applying the convolution formula of the fast algorithm and an augmented Vandermonde decomposition of strong Hankel tensors. We prove that if a lower-order Hankel tensor is positive semidefinite, then a higher-order Hankel tensor with the same generating vector has no negative H-eigenvalues, when (i) the lower order is 2, or (ii) the lower order is even and the higher order is its multiple.

The third part is contributed to $\mathcal{M}$-tensors. We attempt to extend the equivalent definitions of nonsingular M-matrices, such as semi-positivity, monotonicity, nonnegative inverse, etc., to the tensor case. Our results show that the semi-positivity is still an equivalent definition of nonsingular $\mathcal{M}$-tensors, while the monotonicity is not. Furthermore, the generalization of the "nonnegative inverse" property inspires the study of multilinear system of equations. We prove the existence and uniqueness of the positive solutions of nonsingular $\mathcal{M}$-equations with positive right-hand sides, and also propose several iterative methods for computing the positive solutions.

We would like to thank our collaborator Prof. Liqun Qi of the Hong Kong Polytechnic University, who leaded us to the research of tensor spectral theory and always encourages us to explore the topic. We would also like to thank Prof. Eric King-wah Chu of Monash University and Prof. Sanzheng Qiao of McMaster University, who read this book carefully and provided feedback during the writing process.

This work was supported by the National Natural Science Foundation of China under Grant 11271084, School of Mathematical Sciences and Key Laboratory of Mathematics for Nonlinear Sciences, Fudan University.

Contents

Preface   v

I   General Theory   1

1   Introduction and Preliminaries   3
   1.1   What Are Tensors?   3
   1.2   Basic Operations   6
   1.3   Tensor Decompositions   8
   1.4   Tensor Eigenvalue Problems 10

2   Generalized Tensor Eigenvalue Problems   11
   2.1   A Unified Framework 11
   2.2   Basic Definitions 13
   2.3   Several Basic Properties 14
       2.3.1   Number of Eigenvalues 14
       2.3.2   Spectral Radius 15
       2.3.3   Diagonalizable Tensor Pairs 15
       2.3.4   Gershgorin Circle Theorem 16
       2.3.5   Backward Error Analysis 19
   2.4   Real Tensor Pairs 20
       2.4.1   The Crawford Number 21
       2.4.2   Symmetric-Definite Tensor Pairs 22
   2.5   Sign-Complex Spectral Radius 26
       2.5.1   Definitions 26
       2.5.2   Collatz-Wielandt Formula 27
       2.5.3   Properties for Single Tensors 29
       2.5.4   The Componentwise Distance to Singularity 31
       2.5.5   Bauer-Fike Theorem 33
   2.6   An Illustrative Example 34

II   Hankel Tensors   37

3   Fast Tensor-Vector Products   39
   3.1   Hankel Tensors 39
   3.2   Exponential Data Fitting 40
       3.2.1   The One-Dimensional Case 40
       3.2.2   The Multidimensional Case 42
   3.3   Anti-Circulant Tensors 45
       3.3.1   Diagonalization 46
       3.3.2   Singular Values 47

3.3.3 Block Tensors 49
3.4 Fast Hankel Tensor-Vector Product 50
3.5 Numerical Examples 53

4 Inheritance Properties   **59**
4.1 Inheritance Properties 59
4.2 The First Inheritance Property of Hankel Tensors 61
    4.2.1 A Convolution Formula 61
    4.2.2 Lower-Order Implies Higher-Order 63
    4.2.3 SOS Decomposition of Strong Hankel Tensors 65
4.3 The Second Inheritance Property of Hankel Tensors 66
    4.3.1 Strong Hankel Tensors 66
    4.3.2 A General Vandermonde Decomposition of Hankel Matrices 68
    4.3.3 An Augmented Vandermonde Decomposition of Hankel
           Tensors 71
    4.3.4 The Second Inheritance Property of Hankel Tensors . . . 75
4.4 The Third Inheritance Property of Hankel Tensors 77

III $\mathcal{M}$-Tensors   **79**

5 Definitions and Basic Properties   **81**
5.1 Preliminaries 81
    5.1.1 Nonnegative Tensor 81
    5.1.2 From M-Matrix to $\mathcal{M}$-Tensor 82
5.2 Spectral Properties of $\mathcal{M}$-Tensors 83
5.3 Semi-Positivity 84
    5.3.1 Definitions 84
    5.3.2 Semi-Positive $\mathcal{Z}$-Tensors 85
    5.3.3 Proof of Theorem 5.7 87
    5.3.4 General $\mathcal{M}$-Tensors 89
5.4 Monotonicity 90
    5.4.1 Definitions 90
    5.4.2 Properties 90
    5.4.3 A Counter Example 93
    5.4.4 A Nontrivial Monotone $\mathcal{Z}$-Tensor 93
5.5 An Extension of $\mathcal{M}$-Tensors 93
5.6 Summation 95

6 Multilinear Systems with $\mathcal{M}$-Tensors   **97**
6.1 Motivations 97
6.2 Triangular Equations 99
6.3 $\mathcal{M}$-Equations and Beyond 102
    6.3.1 $\mathcal{M}$-Equations 102
    6.3.2 Nonpositive Right-Hand Side 104
    6.3.3 Nonhomogeneous Left-Hand Side 105
    6.3.4 Absolute $\mathcal{M}$-Equations 106
    6.3.5 Banded $\mathcal{M}$-Equation 107
6.4 Iterative Methods for $\mathcal{M}$-Equations 108
    6.4.1 The Classical Iterations 109

6.4.2   The Newton Method for Symmetric $\mathcal{M}$-Equations 111
6.4.3   Numerical Tests 112
6.5   Perturbation Analysis of $\mathcal{M}$-Equations 114
6.5.1   Backward Errors of Triangular $\mathcal{M}$-Equations 115
6.5.2   Condition Numbers 116
6.6   Inverse Iteration 118

Bibliography   **125**

Subject Index   **135**

Part I

General Theory

Chapter 1

Introduction and Preliminaries

We first introduce the concepts and sources of tensors in this chapter. Several essential and frequently used operations involving tensors are also included. Furthermore, two basic topics, tensor decompositions and tensor eigenvalue problems, are briefly discussed at the end of this chapter.

1.1 What Are Tensors?

The term *tensor* or *hypermatrix* in this book refers to a multiway array. The number of the dimensions of a tensor is called its *order*, that is, $\mathcal{A} = (a_{i_1 i_2 \ldots i_m})$ is an mth-order tensor. Particularly, a scalar is a 0th-order tensor, a vector is a 1st-order tensor, and a matrix is a 2nd-order tensor. As other mathematical concepts, tensor or hypermatrix is abstracted from real-world phenomena and other scientific theories. Where do the tensors arise? What kinds of properties do we care most? How many different types of tensors do we have? We will briefly answer these questions employing several illustrative examples in this section.

Example 1.1. As we know, a table is one of the most common realizations of a matrix. We can also understand tensors or hypermatrices as complex tables with multivariables. For instance, if we record the scores of 4 students on 3 subjects for both the midterm and final exams, then we can design a 3rd-order tensor $\mathcal{S}$ of size $4 \times 3 \times 2$ whose (i, j, k) entry s_{ijk} denotes the score of the i-th student on the j-th subjects in the k-th exam. This representation is natural and easily understood, thus it is a convenient data structure for construction and query. However, when we need to print the information on a piece of paper, the 3D structure is apparently not suitable for 2D visualization. Thus we need to unfold the cubic tensor into a matrix. The following two different unfoldings of the same tensor both include all the information in the original complex table. We can see from the two tables that their entries are the same up to a permutation. Actually, there are many different ways to unfold a higher-order tensor into a matrix, and the linkages between them are permutations of indices.

Theory and Computation of Tensors.
http://dx.doi.org/10.1016/B978-0-12-803953-3.50001-0

	Sub. 1		Sub. 2		Sub. 3	
	Mid	Final	Mid	Final	Mid	Final
Std. 1	s_{111}	s_{112}	s_{121}	s_{122}	s_{131}	s_{132}
Std. 2	s_{211}	s_{212}	s_{221}	s_{222}	s_{231}	s_{232}
Std. 3	s_{311}	s_{312}	s_{321}	s_{322}	s_{331}	s_{332}
Std. 4	s_{411}	s_{412}	s_{421}	s_{422}	s_{431}	s_{432}

Table 1.1: The first way to print $\mathcal{S}$.

	Mid			Final		
	Sub. 1	Sub. 2	Sub. 3	Sub. 1	Sub. 2	Sub. 3
Std. 1	s_{111}	s_{121}	s_{131}	s_{112}	s_{122}	s_{132}
Std. 2	s_{211}	s_{221}	s_{231}	s_{212}	s_{222}	s_{232}
Std. 3	s_{311}	s_{321}	s_{331}	s_{312}	s_{322}	s_{332}
Std. 4	s_{411}	s_{421}	s_{431}	s_{412}	s_{422}	s_{432}

Table 1.2: The second way to print $\mathcal{S}$.

Example 1.2. Another important realization of tensors are the storage of color images and videos. A black-and-white image can be stored as a greyscale matrix, whose entries are the greyscale values of the corresponding pixels. Color images are often built from several stacked color channels, each of which represents value levels of the given channel. For example, RGB images are composed of three independent channels for red, green, and blue primary color components. We can apply a 3rd-order tensor $\mathcal{P}$ to store an RGB image, whose (i, j, k) entry denotes the value of the k-th channel in the (i, j) position. ($k = 1, 2, 3$ represent the red, green, and blue channel, respectively.) In order to store a color video, we may need an extra index for the time axis. That is, we employ a 4th-order tensor $\mathcal{M} = (m_{ijkt})$, where $\mathcal{M}(:, :, :, t)$ stores the t-th frame of the video as a color image.

Example 1.3. Denote $\mathbf{x} = (x_1, x_2, \ldots, x_n)^{\top} \in \mathbb{R}^n$. As we know, a degree-1 polynomial $p_1(\mathbf{x}) = c_1 x_1 + c_2 x_2 + \cdots + c_n x_n$ can be rewritten into $p_1(\mathbf{x}) = \mathbf{x}^{\top} \mathbf{c}$, where the vector $\mathbf{c} = (c_1, c_2, \ldots, c_n)^{\top}$. Similarly, a degree-2 polynomial $p_2(\mathbf{x}) = \sum_{i,j=1}^{n} c_{ij} x_i x_j$, that is, a quadratic form, can be simplified into $p_2(\mathbf{x}) = \mathbf{x}^{\top} C \mathbf{x}$, where the matrix $C = (c_{ij})$. By analogy, if we denote an mth-order tensor $\mathcal{C} = (c_{i_1 i_2 \ldots i_m})$ and apply a notation, which will be introduced in the next section, then the degree-m homogeneous polynomial

$$p_m(\mathbf{x}) = \sum_{i_1=1}^{n} \sum_{i_2=1}^{n} \cdots \sum_{i_m=1}^{n} c_{i_1 i_2 \ldots i_m} x_{i_1} x_{i_2} \ldots x_{i_m}$$

can be rewritten as

$$p_m(\mathbf{x}) = \mathcal{C} \mathbf{x}^m.$$

Moreover, $\mathbf{x}^{\top} \mathbf{c} = 0$ is often used to denote a hyperplane in $\mathbb{R}^n$. Similarly, $\mathcal{C} \mathbf{x}^m = 0$ can stand for an degree-m hypersurface in $\mathbb{R}^n$. We shall see in Section 1.2 that the normal vector at a point $\mathbf{x}_0$ on this hypersurface is $\mathbf{n}_{\mathbf{x}_0} = \mathcal{C} \mathbf{x}_0^{m-1}$.

Example 1.4. The Taylor expansion is a well-known mathematical tool. The Taylor series of a real or complex-valued function $f(x)$ that is infinitely differentiable at a real or complex number a is the power series

$$f(x) = \sum_{m=0}^{\infty} \frac{1}{m!} f^{(m)}(a)(x-a)^m.$$

A multivariate function $f(x_1, x_2, \ldots, x_n)$ that is infinitely differentiable at a point $(a_1, a_2, \ldots, a_n)$ also has its Taylor expansion

$$f(x_1, \ldots, x_n) = \sum_{i_1=0}^{\infty} \cdots \sum_{i_n=0}^{\infty} \frac{(x_1-a_1)^{i_1} \ldots (x_n-a_n)^{i_n}}{i_1! \cdots i_n!} \frac{\partial^{i_1+\cdots+i_n} f(a_1, \ldots, a_n)}{\partial x_1^{i_1} \ldots \partial x_n^{i_n}},$$

which is equivalent to

$$f(x_1, \ldots, x_n) = f(a_1, \ldots, a_n) + \sum_{i=1}^{n} \frac{\partial f(a_1, \ldots, a_n)}{\partial x_i}(x_i - a_i)$$

$$+ \frac{1}{2!} \sum_{i=1}^{n} \sum_{j=1}^{n} \frac{\partial^2 f(a_1, \ldots, a_n)}{\partial x_i \partial x_j}(x_i - a_i)(x_j - a_j)$$

$$+ \frac{1}{3!} \sum_{i=1}^{n} \sum_{j=1}^{n} \sum_{k=1}^{n} \frac{\partial^3 f(a_1, \ldots, a_n)}{\partial x_i \partial x_j \partial x_k}(x_i - a_i)(x_j - a_j)(x_k - a_k) + \cdots.$$

Denoting $\mathbf{x} = (x_1, x_2, \ldots, x_n)^\top$ and $\mathbf{a} = (a_1, a_2, \ldots, a_n)^\top$, then we can rewrite the second and the third terms in the above equation as

$$(\mathbf{x} - \mathbf{a})^\top \nabla f(\mathbf{a}) \quad \text{and} \quad (\mathbf{x} - \mathbf{a})^\top \nabla^2 f(\mathbf{a})(\mathbf{x} - \mathbf{a}),$$

where $\nabla f(\mathbf{a})$ and $\nabla^2 f(\mathbf{a})$ are the gradient and the Hessian of $f(\mathbf{x})$ at $\mathbf{a}$, respectively. If we define the mth-order gradient tensor $\nabla^m f(\mathbf{a})$ of $f(\mathbf{x})$ at $\mathbf{a}$ as

$$(\nabla^m f(\mathbf{a}))_{i_1 i_2 \ldots i_m} = \frac{\partial^m f(a_1, a_2, \ldots, a_n)}{\partial x_{i_1} \partial x_{i_2} \ldots \partial x_{i_m}},$$

then the Taylor expansion of a multivariate function can also be expressed by

$$f(\mathbf{x}) = \sum_{m=0}^{\infty} \frac{1}{m!} \nabla^m f(\mathbf{a})(\mathbf{x} - \mathbf{a})^m.$$

Example 1.5. The discrete-time Markov chain is one of the most important models of random processes [10, 28, 81, 114], which assumes that the future is dependent solely on the finite past. This model is so simple and natural that Markov chains are widely employed in many disciplines, such as thermodynamics, statistical mechanics, queueing theory, web analysis, economics, and finance. An sth-order Markov chain is a stochastic process with the Markov property, that is, a sequence of variables $\{Y_t\}_{t=1}^{\infty}$ satisfying

$$\Pr(Y_t = i_1 \mid Y_{t-1} = i_2, \ldots, Y_1 = i_t) = \Pr(Y_t = i_1 \mid Y_{t-1} = i_2, \ldots, Y_{t-s} = i_{s+1}),$$

for all $t > s$. That is, any state depends solely on the immediate past s states. Particularly, when the step length $s = 1$, the sequence $\{Y_t\}_{t=1}^{\infty}$ is a standard first-order Markov chain. Define the transition probability matrix P of an second-order Markov chain as

$$p_{ij} = \Pr(Y_t = i \mid Y_{t-1} = j),$$

which is a stochastic matrix, that is, $p_{ij} \geq 0$ for all $i, j = 1, 2, \ldots, n$ and $\sum_{i=1}^{n} p_{ij} = 1$ for all $j = 1, 2, \ldots, n$. The probability distribution of $\widetilde{Y}_t$ is denoted by a vector

$$(\mathbf{x}_t)_i = \Pr(Y_t = i).$$

Then the Markov chain is modeled by $\mathbf{x}_{t+1} = P\mathbf{x}_t$, thus the stationary probability distribution $\mathbf{x}$ satisfies $\mathbf{x} = P\mathbf{x}$ and is exactly an eigenvector of the transition probability matrix corresponding to the eigenvalue 1.

For higher-order Markov chains, we have similar formulations. Take a second-order Markov chain, that is, $s = 2$, as an example. Define the transition probability tensor $\mathcal{P}$ of a second-order Markov chain as

$$p_{ijk} = \Pr(Y_t = i \,|\, Y_{t-1} = j, Y_{t-2} = k),$$

which is a stochastic tensor, that is, $\mathcal{P} \geq 0$ and $\sum_{i=1}^{n} p_{ijk} = 1$ for all $j, k = 1, 2, \ldots, n$. The probability distribution of $\widetilde{Y}_t$ in the product space can be reshaped into a matrix

$$(X_t)_{i,j} = \Pr(Y_t = i, Y_{t-1} = j).$$

Then the stationary probability distribution in the product space X satisfies

$$X(i,j) = \sum_{k=1}^{n} P(i,j,k) \cdot X(j,k)$$

for all $i, j = 1, 2, \ldots, n$. If we further assume that

$$\Pr(Y_t = i, Y_{t-1} = j) = \Pr(Y_t = i) \cdot \Pr(Y_{t-1} = j)$$

for all $i, j = 1, 2, \ldots, n$ and denote $(\mathbf{x}_t)_i = \Pr(Y_t = i)$, that is, $X_t = \mathbf{x}_t \mathbf{x}_t^{\top}$, then the stationary probability distribution $\mathbf{x}$ satisfies $\mathbf{x} = \mathcal{P}\mathbf{x}^2$ [74]. We shall see in Section 1.2 that $\mathbf{x}$ is a special eigenvector of the tensor $\mathcal{P}$.

From the above examples, we can gain some basic ideas about what tensors are and where they come from. Generally speaking, there are two kinds of tensors: the first kind is a data structure, which admits different dimensions according to the complexity of the data; the second kind is an operator, where it possesses different meanings in different situations.

1.2   Basic Operations

We first introduce several basic tensor operations that will be frequently referred to in the book. One of the difficulties of tensor research is the complicated indices. Therefore we often use some small-size examples rather than exact definitions in this section to describe those essential concepts more clearly. For more detailed definitions, we refer to Chapter 12 in [49] and the references [18, 66, 98].

- To treat or visualize the multidimensional structures, we often reshape a higher-order tensor into a vector or a matrix, which are more familiar to us. The vectorization operator $\text{vec}(\cdot)$ turns tensors into column vectors. Take a $2 \times 2 \times 2$ tensor $\mathcal{A} = (a_{ijk})_{i,j,k=1}^{2}$ for example, then

$$\text{vec}(\mathcal{A}) = (a_{111}, a_{211}, a_{121}, a_{221}, a_{112}, a_{212}, a_{122}, a_{222})^{\top}.$$

There are a lot of different ways to reshape tensors into matrices, which are often referred to as "unfoldings." The most frequently applied one is call the *modal unfolding*. The mode-k unfolding $\mathcal{A}_{(k)}$ of an mth-order tensor $\mathcal{A}$ of size $n_1 \times n_2 \times \cdots \times n_m$ is an n_k-by-(N/n_k) matrix, where $N = n_1 n_2 \ldots n_m$. Again, use the above $2 \times 2 \times 2$ example. Its mode-1, mode-2, and mode-3 unfoldings are

$$\mathcal{A}_{(1)} = \begin{pmatrix} a_{111} & a_{121} & a_{112} & a_{122} \\ a_{211} & a_{221} & a_{212} & a_{222} \end{pmatrix},$$

$$\mathcal{A}_{(2)} = \begin{pmatrix} a_{111} & a_{211} & a_{112} & a_{212} \\ a_{121} & a_{221} & a_{122} & a_{222} \end{pmatrix},$$

$$\mathcal{A}_{(3)} = \begin{pmatrix} a_{111} & a_{211} & a_{121} & a_{221} \\ a_{112} & a_{212} & a_{122} & a_{222} \end{pmatrix},$$

respectively. Sometimes the mode-k unfolding is also denoted as $\text{Unfold}_k(\cdot)$.

- The transposition operation of a matrix is understood as the exchange of the two indices. But higher-order tensors have more indices, thus we have much more transpositions of tensors. If $\mathcal{A}$ is a 3rd-order tensor, then there are six possible transpositions denoted as $\mathcal{A}^{<[\sigma(1),\sigma(2),\sigma(3)]>}$, where $(\sigma(1), \sigma(2), \sigma(3))$ is any of the six permutations of $(1, 2, 3)$. When $\mathcal{B} = \mathcal{A}^{<[\sigma(1),\sigma(2),\sigma(3)]>}$, it means

$$b_{i_{\sigma(1)} i_{\sigma(2)} i_{\sigma(3)}} = a_{i_1 i_2 i_3}.$$

If all the entries of a tensor are invariant under any permutations of the indices, then we call it a *symmetric* tensor. For example, a 3rd-order tensor is said to be symmetric if and only if

$$\mathcal{A}^{<[1,2,3]>} = \mathcal{A}^{<[1,3,2]>} = \mathcal{A}^{<[2,1,3]>} = \mathcal{A}^{<[2,3,1]>} = \mathcal{A}^{<[3,1,2]>} = \mathcal{A}^{<[3,2,1]>}.$$

- Modal tensor-matrix multiplications are essential in this book, which are generalizations of matrix-matrix multiplications. Let $\mathcal{A}$ be an mth-order tensor of size $n_1 \times n_2 \times \cdots \times n_m$ and M be a matrix of size $n_k \times n'_k$, then the mode-k product $\mathcal{A} \times_k M$ of the tensor $\mathcal{A}$ and the matrix M is another m^{th}-order tensor of size $n_1 \times \cdots \times n'_k \times \cdots \times n_m$ with

$$(\mathcal{A} \times_k M)_{i_1 \ldots i_{k-1} j_k i_{k+1} \ldots i_m} = \sum_{i_k=1}^{n_k} a_{i_1 \ldots i_{k-1} i_k i_{k+1} \ldots i_m} \cdot m_{i_k j_k}.$$

Particularly, if A, M_1 and M_2 are all matrices, then $A \times_1 M_1 \times_2 M_2 = M_1^{\top} A M_2$. Easily verified, the tensor-matrix multiplications satisfy that

1. $A \times_k M_k \times_l M_l = A \times_l M_l \times_k M_k$, if $k \neq l$,
2. $A \times_k M_1 \times_k M_2 = A \times_k (M_1 M_2)$,

3. $\mathcal{A} \times_k (\alpha_1 M_1 + \alpha_2 M_2) = \alpha_1 \mathcal{A} \times_k M_1 + \alpha_1 \mathcal{A} \times_k M_1$,

4. $\mathrm{Unfold}_k(\mathcal{A} \times_k M) = M^\top \mathcal{A}_{(k)}$,

5. $\mathrm{vec}(\mathcal{A} \times_1 M_1 \times_2 M_2 \cdots \times_m M_m) = (M_m \otimes \cdots \otimes M_2 \otimes M_1)^\top \mathrm{vec}(\mathcal{A})$,

where $\mathcal{A}$ is a tensor, M_k are matrices, and α_1, α_2 are scalars.

- If the matrices degrade into column vectors, then we obtain another cluster of important notations for tensor spectral theory. Let $\mathcal{A}$ be an mth-order n-dimensional tensor, that is, of size $n \times n \times \cdots \times n$, and $\mathbf{x}$ be a vector of length n, then for simplicity:

$$\mathcal{A}\mathbf{x}^m = \mathcal{A} \times_1 \mathbf{x} \times_2 \mathbf{x} \times_3 \mathbf{x} \cdots \times_m \mathbf{x} \text{ is a scalar,}$$
$$\mathcal{A}\mathbf{x}^{m-1} = \mathcal{A} \qquad \times_2 \mathbf{x} \times_3 \mathbf{x} \cdots \times_m \mathbf{x} \text{ is a vector,}$$
$$\mathcal{A}\mathbf{x}^{m-2} = \mathcal{A} \qquad\qquad \times_3 \mathbf{x} \cdots \times_m \mathbf{x} \text{ is a matrix.}$$

- Like the vector case, an inner product of two tensors $\mathcal{A}$ and $\mathcal{B}$ of the same size are defined by

$$\langle \mathcal{A}, \mathcal{B} \rangle = \sum_{i_1=1}^{n_1} \cdots \sum_{i_m=1}^{n_m} a_{i_1 i_2 \ldots i_m} \cdot b_{i_1 i_2 \ldots i_m},$$

which is exactly the usual inner product of the two vectors $\mathrm{vec}(\mathcal{A})$ and $\mathrm{vec}(\mathcal{B})$.

- The outer product of two tensors is a higher-order tensor. Let $\mathcal{A}$ and $\mathcal{B}$ be m^{th}-order and $(m')^{\mathrm{th}}$-order tensors, respectively. Then their outer product $\mathcal{A} \circ \mathcal{B}$ is an $(m + m')^{\mathrm{th}}$-order tensor with

$$(\mathcal{A} \circ \mathcal{B})_{i_1 \ldots i_m j_1 \ldots j_{m'}} = a_{i_1 i_2 \ldots i_m} \cdot b_{j_1 j_2 \ldots j_{m'}}.$$

If $\mathbf{a}$ and $\mathbf{b}$ are vectors, then $\mathbf{a} \circ \mathbf{b} = \mathbf{a}\mathbf{b}^\top$.

- We sometimes refer to the Hadamard product of two tensors with the same size as

$$(\mathcal{A} \underset{\mathrm{HAD}}{\otimes} \mathcal{B})_{i_1 i_2 \ldots i_m} = a_{i_1 i_2 \ldots i_m} \cdot b_{i_1 i_2 \ldots i_m}.$$

The Hadamard product will also be denoted as $\mathcal{A} \cdot * \mathcal{B}$ in the descriptions of some algorithms, which is a MATLAB-type notation.

1.3   Tensor Decompositions

Given a tensor, how can we retrieve the information hidden inside? One reasonable answer is the tensor decomposition approach. Existing tensor decompositions include the Tucker-type decompositions, the CANDECOMP/PARAFAC (CP) decomposition, tensor train representation, etc. For those readers interested in tensor decompositions, we recommend the survey papers [50, 66]. Some tensor decompositions are generalizations of the singular value decomposition (SVD) [49].

The SVD is one of the most important tools for matrix analysis and computation. Any matrix $A \in \mathbb{R}^{m \times n}$ with rank r has the decomposition

$$A = U \Sigma V^\top,$$

where $U \in \mathbb{R}^{m \times r}$ and $V \in \mathbb{R}^{n \times r}$ are column orthogonal matrices, that is, $U^\top U = I_m$ and $V^\top V = I_n$, and $\Sigma = \text{diag}(\sigma_1, \sigma_2, \ldots, \sigma_r)$ is a positive diagonal matrix. Then σ_k are called the *singular values* of A. If $U = [\mathbf{u}_1, \mathbf{u}_2, \ldots, \mathbf{u}_r]$ and $V = [\mathbf{v}_1, \mathbf{v}_2, \ldots, \mathbf{v}_r]$, then the SVD can be rewritten asinto

$$A = \sum_{i=1}^{r} \sigma_i \mathbf{u}_i \mathbf{v}_i^\top,$$

which represents the matrix A as a sum of several rank-one matrices. For an arbitrary nonsingular matrix $M \in \mathbb{R}^{r \times r}$, denote $L = U\Sigma M^\top \in \mathbb{R}^{m \times r}$ and $R = VM^{-1} \in \mathbb{R}^{n \times r}$. Then

$$A = LR^\top$$

is a low-rank decomposition if r is much smaller than m and n. These three equivalent formulas of the SVD are all extended to the higher-order tensors.

Let $\mathcal{A}$ be an mth-order tensor of size $n_1 \times n_2 \times \cdots \times n_m$. The *Tucker-type decompositions* of the tensor $\mathcal{A}$ has the form

$$\mathcal{A} = \mathcal{S} \times_1 U_1^\top \times_2 U_2^\top \cdots \times_m U_m^\top,$$

where $\mathcal{S}$ is an mth-order tensor of size $r \times r \times \cdots \times r$, called the *core tensor*, and $U_k \in \mathbb{R}^{n \times r}$ for $k = 1, 2, \ldots, m$. Note that the first formula $A = U\Sigma V^\top$ of the SVD can be rewritten into this form $A = \Sigma \times_1 U^\top \times_2 V^\top$. If there are no restrictions on $\mathcal{S}$ and U_k, then we have an infinite number of Tucker-type decompositions, most of which are no more informative. The *higher-order singular value decomposition* (HOSVD), proposed in [33, 35], is a special Tucker-type decomposition, which satisfies that

- the core $\mathcal{S}$ is *all-orthogonal* ($\langle \mathcal{S}_{i_k=\alpha}, \mathcal{S}_{i_k=\beta} \rangle = 0$ for all $k = 1, 2, \ldots, m$, $\alpha \neq \beta$) and *ordered* ($\|\mathcal{S}_{i_k=1}\|_F \geq \|\mathcal{S}_{i_k=2}\|_F \geq \cdots \geq \|\mathcal{S}_{i_k=n_k}\|_F$ for all $k = 1, 2, \ldots, m$); and

- the matrices U_k are column orthogonal.

Actually, the matrix U_k consists of the left singular vectors of the mode-k unfolding $\mathcal{A}_{(k)}$, and $\|\mathcal{S}_{i_k=1}\|_F, \|\mathcal{S}_{i_k=2}\|_F, \ldots, \|\mathcal{S}_{i_k=n_k}\|_F$ are exactly the singular values of $\mathcal{A}_{(k)}$. One can easily verify that if a matrix is all-orthogonal then it must be a diagonal matrix. Thus the 2nd-order version of the HOSVD is exactly the SVD.

Nevertheless, the core tensor $\mathcal{S}$ of the HOSVD is never sparse in the higher-order cases, thus complicated and obscure. If we restrict the core tensor to be diagonal, that is, the entries except for $s_{ii\ldots i}$ $(i = 1, 2, \ldots, r)$ are all zero, then we may relax the restriction on the number of terms r. Denote the diagonal entries of $\mathcal{S}$ as $\sigma_1, \sigma_2, \ldots, \sigma_r$, and $U_k = [\mathbf{u}_{k1}, \mathbf{u}_{k2}, \ldots, \mathbf{u}_{kr}]$ for all $k = 1, 2, \ldots, m$. The *CP decomposition* of the tensor $\mathcal{A}$ is referred to as

$$\mathcal{A} = \sum_{i=1}^{r} \sigma_i \mathbf{u}_{1i} \circ \mathbf{u}_{2i} \circ \cdots \circ \mathbf{u}_{mi},$$

where r might be larger than n. Each term $\mathbf{u}_{1i} \circ \mathbf{u}_{2i} \circ \cdots \circ \mathbf{u}_{mi}$ in the CP decomposition is called a *rank-one* tensor. The least number of the terms, that is,

$$R = \min \left\{ r : \mathcal{A} = \sum_{i=1}^{r} \sigma_i \mathbf{u}_{1i} \circ \mathbf{u}_{2i} \circ \cdots \circ \mathbf{u}_{mi} \right\},$$

is called the *CP-rank* of the tensor $\mathcal{A}$. The computation or estimation of the CP-ranks of higher-order tensors is NP-hard [54], and still a hard task in tensor research.

The *tensor train format* (TT) [92, 91] for higher-order tensors extends the low-rank decomposition of matrices. Let $\mathcal{A}$ be an mth-order tensor. Then it can be expressed in the following format

$$a_{i_1 i_2 \ldots i_m} = \sum_{k_1=1}^{r_1} \cdots \sum_{k_{m-1}=1}^{r_{m-1}} (\mathcal{G}_1)_{i_1 k_1} (\mathcal{G}_2)_{k_1 i_2 k_2} \cdots (\mathcal{G}_{m-1})_{k_{m-2} i_{m-1} k_{m-1}} (\mathcal{G}_m)_{k_{m-1} i_m}$$

for all $i_1, i_2, \ldots, i_m$, where $\mathcal{G}_1$ and $\mathcal{G}_m$ are matrices and $\mathcal{G}_2, \ldots, \mathcal{G}_{m-1}$ are 3rd-order tensors. If $r_1, r_2, \ldots, r_{m-1}$ are much smaller than the tensor size, then the TT representation will greatly reduce the storage cost for the tensor.

1.4 Tensor Eigenvalue Problems

We have introduced the tensor-vector products in the previous sections. Given an mth-order n-dimensional square tensor $\mathcal{A}$, notice that $\mathbf{x} \mapsto \mathcal{A}\mathbf{x}^{m-1}$ is a nonlinear operator from $\mathbb{C}^n$ to itself. Thus we should consider some characteristic values of this operator. For this sake, we can define several tensor eigenvalue problems.

First, we introduce a homogeneous tensor eigenvalue problem. If a scalar $\lambda \in \mathbb{C}$ and a nonzero vector $\mathbf{x} \in \mathbb{C}^n$ satisfy

$$\mathcal{A}\mathbf{x}^{m-1} = \lambda \mathbf{x}^{[m-1]},$$

where $\mathbf{x}^{[m-1]} = [x_1^{m-1}, x_2^{m-1}, \ldots, x_n^{m-1}]^{\top}$, then λ is called an *eigenvalue* of $\mathcal{A}$ and $\mathbf{x}$ is called a corresponding *eigenvector*. Furthermore, when the tensor $\mathcal{A}$, the scalar λ and the vector $\mathbf{x}$ are all real, we call λ an *H-eigenvalue* of $\mathcal{A}$ and $\mathbf{x}$ a corresponding *H-eigenvector*. For a real symmetric tensor $\mathcal{A}$, its H-eigenvectors are exactly the KKT points of the polynomial optimization problem

$$\begin{aligned} \max/\min \quad & \mathcal{A}\mathbf{x}^m, \\ \text{s.t.} \quad & x_1^m + x_2^m + \cdots + x_n^m = 1. \end{aligned}$$

This was the original motivation when Qi [98] and Lim [76] first introduced this kind of eigenvalue problem.

We also have other nonhomogeneous tensor eigenvalue definitions. For example, if a scalar $\lambda \in \mathbb{C}$ and a nonzero vector $\mathbf{x} \in \mathbb{C}^n$ satisfy

$$\mathcal{A}\mathbf{x}^{m-1} = \lambda \mathbf{x} \quad \text{with } \mathbf{x}^{\top}\mathbf{x} = 1,$$

then λ is called an *E-eigenvalue* of $\mathcal{A}$ and $\mathbf{x}$ is called a corresponding *E-eigenvector* [99]. Furthermore, when the tensor $\mathcal{A}$, the scalar λ, and the vector $\mathbf{x}$ are all real, we call λ an *Z-eigenvalue* of $\mathcal{A}$ and $\mathbf{x}$ a corresponding *Z-eigenvector*. For a real symmetric tensor $\mathcal{A}$, its Z-eigenvectors are exactly the KKT points of the polynomial optimization problem

$$\begin{aligned} \max/\min \quad & \mathcal{A}\mathbf{x}^m, \\ \text{s.t.} \quad & x_1^2 + x_2^2 + \cdots + x_n^2 = 1. \end{aligned}$$

We will introduce more definitions of tensor eigenvalues in Chapter 2, which can be unified into a *generalized tensor eigenvalue* problem $\mathcal{A}\mathbf{x}^{m-1} = \lambda \mathcal{B}\mathbf{x}^{m-1}$. Further discussions will be conducted on this unified framework.

Chapter 2

Generalized Tensor Eigenvalue Problems

Generalized matrix eigenvalue problems are essential in scientific and engineering computations. The generalized eigenvalue formula $A\mathbf{x} = \lambda B\mathbf{x}$ is also a very pervasive model, which covers the matrix polynomial eigenvalue problems. There has been extensive study of both the theories and the computations of generalized matrix eigenvalue problems, and one can refer to [6, 49, 113] for more details. Applying the notations introduced in Chapter 1, the generalized tensor eigenvalue problem is referred to finding λ and $\mathbf{x}$ satisfying

$$\mathcal{A}\mathbf{x}^{m-1} = \lambda \mathcal{B}\mathbf{x}^{m-1}.$$

The detailed definition will be introduced later.

2.1 A Unified Framework

Since Qi [98] and Lim [76] proposed the definition of tensor eigenvalues independently in 2005, a great deal of mathematical effort has been devoted to the study of the eigenproblems of single tensors, especially for the nonnegative tensors [18, 19]. Nevertheless, there has been much less specific research on the generalized eigenproblems, since Chang et al. [19, 20] introduced the generalized tensor eigenvalues. Recently, Kolda and Mayo [68], Cui et al. [30], and Chen et al. [25] proposed numerical algorithms for generalized tensor eigenproblems. Kolda and Mayo [68] applied an adaptive shifted power method by solving the optimization problem

$$\max_{\|\mathbf{x}\|=1} \frac{\mathcal{A}\mathbf{x}^m}{\mathcal{B}\mathbf{x}^m} \|\mathbf{x}\|^m.$$

In Cui, Dai, and Nie's paper [30], they impose the constraint $\lambda_{k+1} < \lambda_k - \delta$ when the k-th eigenvalue is obtained, so that the $(k+1)$-th eigenvalue can also be computed by the same method as the previous ones. Chen, Han, and Zhou [25] proposed homotopy methods for generalized tensor eigenvalue problems.

It was pointed out in [19, 20, 30, 68] that the generalized eigenvalue framework unifies several definitions of tensor eigenvalues, such as the eigenvalues and

Theory and Computation of Tensors.
http://dx.doi.org/10.1016/B978-0-12-803953-3.50002-2

the H-eigenvalues [98], that is,

$$\mathcal{A}\mathbf{x}^{m-1} = \lambda \mathbf{x}^{[m-1]}, \text{ where } \mathbf{x}^{[m-1]} = [x_1^{m-1}, x_2^{m-1}, \ldots, x_n^{m-1}]^\top,$$

the E-eigenvalues and the Z-eigenvalues [99], that is,

$$\mathcal{A}\mathbf{x}^{m-1} = \lambda \mathbf{x} \text{ with } \mathbf{x}^\top \mathbf{x} = 1,$$

and the D-eigenvalues [103], that is,

$$\mathcal{A}\mathbf{x}^{m-1} = \lambda D\mathbf{x} \text{ with } \mathbf{x}^\top D\mathbf{x} = 1,$$

where D is a positive definite matrix. We shall present several other potential applications of generalized tensor eigenvalue problems.

The first example is the higher-order Markov chain [28]. Li and Ng [74] reformulated an $(m-1)^{\text{st}}$-order Markov chain problem into searching for a nonnegative vector $\tilde{\mathbf{x}}$ satisfying $\mathcal{P}\tilde{\mathbf{x}}^{m-1} = \tilde{\mathbf{x}}$ with $\tilde{x}_1 + \tilde{x}_2 + \cdots + \tilde{x}_n = 1$, where $\mathcal{P}$ is the transition probability tensor of the Markov chain. Let $\mathcal{E}$ be a tensor such that

$$(\mathcal{E}\mathbf{x}^{m-1})_i = x_i(x_1 + x_2 + \cdots + x_n)^{m-2} \quad (i = 1, 2, \ldots, n)$$

for an arbitrary vector $\mathbf{x} = (x_1, x_2, \ldots, x_n)^\top$. Thus a nonnegative vector $\tilde{\mathbf{x}}$ that satisfies the homogenous equality $\mathcal{P}\tilde{\mathbf{x}}^{m-1} = \mathcal{E}\tilde{\mathbf{x}}^{m-1}$ is exactly what is required.

The US-eigenvalue [85] from quantum information processing is another special case of the generalized tensor eigenvalue. Let $\mathcal{S}$ be an mth-order n-dimensional symmetric tensor. Then the US-eigenvalues are defined by

$$\left\{ \begin{array}{l} \bar{\mathcal{S}}\mathbf{x}^{m-1} = \lambda\bar{\mathbf{x}}, \\ \mathcal{S}\bar{\mathbf{x}}^{m-1} = \lambda\mathbf{x}, \end{array} \right. \quad \text{with} \quad \|\mathbf{x}\|_2 = 1.$$

Denote $\tilde{\mathcal{S}}$ as an mth-order $(2n)$-dimensional tensor with

$$\tilde{\mathcal{S}}(1:n, \ldots, 1:n) = \bar{\mathcal{S}} \quad \text{and} \quad \tilde{\mathcal{S}}(n+1:2n, \ldots, n+1:2n) = \mathcal{S},$$

and $\tilde{\mathbf{x}} = (\mathbf{x}^\top, \bar{\mathbf{x}}^\top)^\top$. When m is even, let $\mathcal{J}$ be a tensor such that

$$(\mathcal{J}\mathbf{y}^{m-1})_i = \left\{ \begin{array}{ll} y_{i+n}(y_1 y_{n+1} + y_2 y_{n+2} + \cdots + y_n y_{2n})^{m/2-1}, & i = 1, 2, \ldots, n, \\ y_{i-n}(y_1 y_{n+1} + y_2 y_{n+2} + \cdots + y_n y_{2n})^{m/2-1}, & i = n+1, \ldots, 2n, \end{array} \right.$$

for an arbitrary vector $\mathbf{y} \in \mathbb{C}^{2n}$. Noticing that $\tilde{x}_1\tilde{x}_{n+1} + \tilde{x}_2\tilde{x}_{n+2} + \cdots + \tilde{x}_n\tilde{x}_{2n} = 1$, we rewrite the definition of the US-eigenvalue into $\tilde{\mathcal{S}}\tilde{\mathbf{x}}^{m-1} = \lambda\mathcal{J}\tilde{\mathbf{x}}^{m-1}$.

Another potential application is from multilabel learning, where the hypergraphs are naturally involved [116]. The *Laplacian tensor* of a hypergraph is widely studied in [29, 58, 73]. Denote $\mathcal{L}$ as the Laplacian tensor of the hypergraph induced by the classification structures. Then $\mathcal{L}\mathbf{x}^m$ presents the clustering score of the data points $\{x_1, x_2, \ldots, x_n\}$ when m is even. Borrowing the idea of graph embedding [127], we can derive a framework of multilabel learning

$$\begin{array}{ll} \max & \mathcal{L}\mathbf{x}^m, \\ \text{s.t.} & \mathcal{L}_p\mathbf{x}^m = 1, \end{array}$$

where $\mathcal{L}_p$ is the Laplacian tensor of a penalty hypergraph, which removes some trivial relations. Employing the method of Lagrange multipliers, we can transform the optimization problem into a generalized tensor eigenvalue problem $\mathcal{L}\mathbf{x}^{m-1} = \lambda \mathcal{L}_p \mathbf{x}^{m-1}$.

Moreover, consider a homogenous polynomial dynamical system [43],

$$\frac{d\mathcal{B}\mathbf{u}(t)^{m-1}}{dt} = \mathcal{A}\mathbf{u}(t)^{m-1}.$$

We explored the stability of the above system in [43]. Similarly to the linear case, if we require a solution which has the form $\mathbf{u}(t) = \mathbf{x} \cdot e^{\lambda t}$, then a generalized tensor eigenvalue problem $\mathcal{A}\mathbf{x}^{m-1} = (m-1)\lambda\mathcal{B}\mathbf{x}^{m-1}$ is naturally raised from this homogenous system.

The generalized tensor eigenvalue problem is an interesting topic, although its computational solutions are not well-developed so far. Therefore we aim at investigating the generalized tensor eigenvalue problems theoretically, especially on some perturbation properties as a prerequisite of the numerical computations in [68].

2.2 Basic Definitions

We will first introduce some concepts and notations involved. Let $\mathbb{K}_{1,2}$ denote a projective plane [113], in which $(\alpha_1, \beta_1), (\alpha_2, \beta_2) \in \mathbb{K} \times \mathbb{K}$ are regarded as the same point, if there is a nonzero scalar $\gamma \in \mathbb{K}$ such that $(\alpha_1, \beta_1) = (\gamma\alpha_2, \gamma\beta_2)$. We can take $\mathbb{K}$ as the complex number field $\mathbb{C}$, the real number field $\mathbb{R}$, or the nonnegative number cone $\mathbb{R}_+$.

The determinant of a tensor is investigated by Qi et al. [98, 57]. The determinant of an mth-order n-dimensional tensor $\mathcal{A}$ is the resultant [31] of the system of homogeneous equations $\mathcal{A}\mathbf{x}^{m-1} = \mathbf{0}$, which is also the unique polynomial on the entries of $\mathcal{A}$ satisfying that

- $\det(\mathcal{A}) = 0$ if and only if $\mathcal{A}\mathbf{x}^{m-1} = \mathbf{0}$ has a nonzero solution;

- $\det(\mathcal{I}) = 1$, where $\mathcal{I}$ is the unit tensor;

- $\det(\mathcal{A})$ is an irreducible polynomial on the entries of $\mathcal{A}$.

Furthermore, $\det(\mathcal{A})$ is homogeneous of degree $n(m-1)^{n-1}$. As a simple example in [98], the determinant of a $2 \times 2 \times 2$ tensor is defined by

$$\det(\mathcal{A}) = \det\begin{pmatrix} a_{111} & a_{112} + a_{121} & a_{122} & 0 \\ 0 & a_{111} & a_{112} + a_{121} & a_{122} \\ a_{211} & a_{212} + a_{221} & a_{222} & 0 \\ 0 & a_{211} & a_{212} + a_{221} & a_{222} \end{pmatrix}.$$

Now we can define the generalized eigenvalue problems of tensor pairs which is similar to the matrix case [6, 49, 113]. Let $\mathcal{A}$ and $\mathcal{B}$ be two m^{th}-order tensors in $\mathbb{C}^{n \times n \times \cdots \times n}$. We call $\{\mathcal{A}, \mathcal{B}\}$ a *regular tensor pair*, if

$$\det(\beta\mathcal{A} - \alpha\mathcal{B}) \neq 0 \text{ for some } (\alpha, \beta) \in \mathbb{C}_{1,2}.$$

Reversely, we call $\{\mathcal{A}, \mathcal{B}\}$ a *singular tensor pair*, if

$$\det(\beta\mathcal{A} - \alpha\mathcal{B}) = 0 \text{ for all } (\alpha, \beta) \in \mathbb{C}_{1,2}.$$

In this chapter, we focus on the regular tensor pairs.

If $\{\mathcal{A}, \mathcal{B}\}$ is a regular tensor pair and there exists $(\alpha, \beta) \in \mathbb{C}_{1,2}$ with a nonzero vector $\mathbf{x} \in \mathbb{C}^n$ such that

$$(2.1) \qquad\qquad \beta \mathcal{A} \mathbf{x}^{m-1} = \alpha \mathcal{B} \mathbf{x}^{m-1},$$

then we call (α, β) an *eigenvalue* of the regular tensor pair $\{\mathcal{A}, \mathcal{B}\}$ and $\mathbf{x}$ the corresponding *eigenvector*. When $\mathcal{B}$ is nonsingular, that is, $\det(\mathcal{B}) \neq 0$, no nonzero vector $\mathbf{x} \in \mathbb{C}^n$ satisfies that $\mathcal{B} \mathbf{x}^{m-1} = \mathbf{0}$, according to [57, Theorem 3.1]. Thus $\beta \neq 0$ if (α, β) is an eigenvalue of $\{\mathcal{A}, \mathcal{B}\}$. We also call $\lambda = \alpha/\beta \in \mathbb{C}$ an eigenvalue of the tensor pair $\{\mathcal{A}, \mathcal{B}\}$ when $\det(\mathcal{B}) \neq 0$. Denote the *spectrum*, that is, the set of all the eigenvalues, of $\{\mathcal{A}, \mathcal{B}\}$ as

$$\lambda(\mathcal{A}, \mathcal{B}) = \{(\alpha, \beta) \in \mathbb{C}_{1,2} : \det(\beta \mathcal{A} - \alpha \mathcal{B}) = 0\},$$

or, for a nonsingular $\mathcal{B}$, we also denote

$$\lambda(\mathcal{A}, \mathcal{B}) = \{\lambda \in \mathbb{C} : \det(\mathcal{A} - \lambda \mathcal{B}) = 0\}.$$

This chapter is devoted to the properties and perturbations of the spectrum $\lambda(\mathcal{A}, \mathcal{B})$ for a regular tensor pair $\{\mathcal{A}, \mathcal{B}\}$.

2.3   Several Basic Properties

There are plenty of theoretical results about the regular matrix pairs [6, 49, 113]. In this section, we extend several of them to the regular tensor pairs. Note that some of the generalized results maintain the forms similar to the matrix case, while others do not.

2.3.1   Number of Eigenvalues

How many eigenvalues does a regular tensor pair have? This might be the first question about the spectrum of a regular tensor pair. We have the following result:

Theorem 2.1. *The number of the eigenvalues of an mth-order n-dimensional regular tensor pair is $n(m-1)^{n-1}$ counting multiplicity.*

Proof. If $\{\mathcal{A}, \mathcal{B}\}$ is an mth-order n-dimensional regular tensor pair, then there exists such $(\gamma, \delta) \in \mathbb{C}_{1,2}$ that $\det(\delta \mathcal{A} - \gamma \mathcal{B}) \neq 0$. Assume that (γ, δ) is normalized to $|\gamma|^2 + |\delta|^2 = 1$. We have the transformation, as the matrix case [115, Page 300],

$$\begin{cases} \widetilde{\mathcal{A}} = \bar{\gamma} \mathcal{A} + \bar{\delta} \mathcal{B}, \\ \widetilde{\mathcal{B}} = \delta \mathcal{A} - \gamma \mathcal{B}, \end{cases} \quad \Leftrightarrow \quad \begin{cases} \mathcal{A} = \gamma \widetilde{\mathcal{A}} + \bar{\delta} \widetilde{\mathcal{B}}, \\ \mathcal{B} = \delta \widetilde{\mathcal{A}} - \bar{\gamma} \widetilde{\mathcal{B}}. \end{cases}$$

Thus there is a one-to-one map between $\lambda(\mathcal{A}, \mathcal{B})$ and $\lambda(\widetilde{\mathcal{A}}, \widetilde{\mathcal{B}})$:

$$\begin{cases} \widetilde{\alpha} = \alpha \bar{\gamma} + \beta \bar{\delta}, \\ \widetilde{\beta} = \alpha \delta - \beta \gamma, \end{cases} \quad \Leftrightarrow \quad \begin{cases} \alpha = \widetilde{\alpha} \gamma + \widetilde{\beta} \bar{\delta}, \\ \beta = \widetilde{\alpha} \delta - \widetilde{\beta} \bar{\gamma}. \end{cases}$$

Since $\det(\widetilde{\mathcal{B}}) \neq 0$, then the eigenvalues of $\{\widetilde{\mathcal{A}}, \widetilde{\mathcal{B}}\}$ are exactly the complex roots of the polynomial equation $\det(\widetilde{\mathcal{A}} - \lambda \widetilde{\mathcal{B}}) = 0$. By [57, Proposition 2.4], the degree of the polynomial $\det(\widetilde{\mathcal{A}} - \lambda\widetilde{\mathcal{B}})$ of λ is no more than $n(m-1)^{n-1}$. We shall show that the degree is exactly $n(m-1)^{n-1}$.

Denote the coefficient of the item $\lambda^{n(m-1)^{n-1}}$ in $\det(\widetilde{\mathcal{A}} - \lambda\widetilde{\mathcal{B}})$ as $f(\widetilde{\mathcal{A}}, \widetilde{\mathcal{B}})$. From the definition of $\det(\widetilde{\mathcal{A}} - \lambda\widetilde{\mathcal{B}})$, we see that $f(\widetilde{\mathcal{A}}, \widetilde{\mathcal{B}})$ is actually independent of the entries of $\widetilde{\mathcal{A}}$. Thus $f(\widetilde{\mathcal{A}}, \widetilde{\mathcal{B}}) = f(\mathcal{O}, \widetilde{\mathcal{B}})$, where $\mathcal{O}$ denotes the zero tensor. It is easy to verify that $f(\mathcal{O}, \widetilde{\mathcal{B}}) = \det(\widetilde{\mathcal{B}}) \neq 0$. Thus $f(\mathcal{I}, \widetilde{\mathcal{B}})$ is nonzero, and so is $f(\widetilde{\mathcal{A}}, \widetilde{\mathcal{B}})$. Therefore the polynomial $\det(\widetilde{\mathcal{A}} - \lambda\widetilde{\mathcal{B}}) = 0$ has $n(m-1)^{n-1}$ roots counting multiplicity. □

2.3.2 Spectral Radius

Spectral radius is an important quantity for the eigenvalue problems. To define the spectral radius of a regular tensor pair, we introduce a representation of the points in $\mathbb{R}_{1,2}$ first. Let (s, c) be a point in $\mathbb{R}_{1,2}$. Without loss of generality, we can assume that $c \geq 0$ and $s > 0$ when $c = 0$. Define an angle in $(-\pi/2, \pi/2]$ as

$$\theta(s, c) := \arcsin \frac{s}{\sqrt{s^2 + c^2}}.$$

When c is fixed, $\theta(\cdot, c)$ is an increasing function; When s is fixed, $\theta(s, \cdot)$ is a decreasing function.

For a regular tensor pair $\{\mathcal{A}, \mathcal{B}\}$, we define the θ-*spectral radius* as

$$\rho_\theta(\mathcal{A}, \mathcal{B}) := \max_{(\alpha,\beta)\in\lambda(\mathcal{A},\mathcal{B})} \theta(|\alpha|, |\beta|).$$

Define another nonnegative function about $\{\mathcal{A}, \mathcal{B}\}$ as

$$\psi_\theta(\mathcal{A}, \mathcal{B}) := \max_{\mathbf{x}\in\mathbb{C}^n\backslash\{\mathbf{0}\}} \theta\big(\|\mathcal{A}\mathbf{x}^{m-1}\|, \|\mathcal{B}\mathbf{x}^{m-1}\|\big).$$

It is apparent that $\rho_\theta(\mathcal{A}, \mathcal{B}) \leq \psi_\theta(\mathcal{A}, \mathcal{B})$, since $\theta\big(\|\mathcal{A}\mathbf{x}^{m-1}\|, \|\mathcal{B}\mathbf{x}^{m-1}\|\big) = \theta(|\alpha|, |\beta|)$, if $(\alpha, \beta) \in \lambda(\mathcal{A}, \mathcal{B})$ and $\mathbf{x}$ is the corresponding eigenvector.

When $\det(\mathcal{B}) \neq 0$, we can define the *spectral radius* in a simpler way

$$\rho(\mathcal{A}, \mathcal{B}) := \max_{\lambda\in\lambda(\mathcal{A},\mathcal{B})} |\lambda|.$$

Similarly, we define a nonnegative function about $\{\mathcal{A}, \mathcal{B}\}$ as

$$\psi(\mathcal{A}, \mathcal{B}) := \max_{\mathbf{x}\in\mathbb{C}^n\backslash\{\mathbf{0}\}} \frac{\|\mathcal{A}\mathbf{x}^{m-1}\|}{\|\mathcal{B}\mathbf{x}^{m-1}\|}.$$

It is easy to verify that $\tan \rho_\theta(\mathcal{A}, \mathcal{B}) = \rho(\mathcal{A}, \mathcal{B}) \leq \psi(\mathcal{A}, \mathcal{B}) = \tan \psi_\theta(\mathcal{A}, \mathcal{B})$. Furthermore, if $\mathcal{B}$ is fixed, then the nonnegative function $\psi(\cdot, \mathcal{B})$ is a seminorm on $\mathbb{C}^{n\times n\times\cdots\times n}$. When $\mathcal{A}, \mathcal{B}$ are matrices and $\mathcal{B} = I$, the result becomes the familiar one that the spectral radius of a matrix is always no larger than its norm.

2.3.3 Diagonalizable Tensor Pairs

Diagonalizable matrix pairs play an important role in the perturbation theory of the generalized eigenvalues [113, Section 6.2.3]. A tensor $\mathcal{A}$ is said to be *diagonal* if all its entries except the diagonal ones $a_{ii...i}$ $(i = 1, 2, \ldots, n)$ are zeros. We call $\{\mathcal{A}, \mathcal{B}\}$ a *diagonalizable tensor pair* if there are two nonsingular matrices P and Q such that

$$\mathcal{C} = P^{-1}\mathcal{A}Q^{m-1} \quad \text{and} \quad \mathcal{D} = P^{-1}\mathcal{B}Q^{m-1}$$

are two diagonal tensors. Let the diagonal entries of $\mathcal{C}$ and $\mathcal{D}$ be $\{c_1, c_2, \ldots, c_n\}$ and $\{d_1, d_2, \ldots, d_n\}$, respectively. If $(c_i, d_i) \neq (0, 0)$ for all $i = 1, 2, \ldots, n$, then $\{\mathcal{A}, \mathcal{B}\}$ is a regular tensor pair. Furthermore, (c_i, d_i) are exactly all the eigenvalues of $\{\mathcal{A}, \mathcal{B}\}$, and their multiplicities are $(m-1)^{n-1}$.

It should be pointed out that the concept of "diagonalizable tensor pair" is not so general as the concept of "diagonalizable matrix pair" [113, Section 6.2.3]. For instance, if matrices A and B are both symmetric, then the matrix pair $\{A, B\}$ must be diagonalizable. However, this is not true for symmetric tensor pairs. Hence we shall present a nontrivial diagonalizable tensor pair to illustrate that it is a reasonable definition.

Example 2.1. Let $\mathcal{A}$ and $\mathcal{B}$ be two m^{th}-order n-dimensional anti-circulant tensors [39]. According to the result in that paper, we know that

$$\mathcal{C} = F_n^*\mathcal{A}(F_n^*)^{m-1} \quad \text{and} \quad \mathcal{D} = F_n^*\mathcal{B}(F_n^*)^{m-1}$$

are both diagonal, where F_n is the n-by-n Fourier matrix [49, Section 1.4]. Therefore an anti-circulant tensor pair must be diagonalizable.

2.3.4   Gershgorin Circle Theorem

The Gershgorin circle theorem [121] is useful to bound the eigenvalues of a matrix. Stewart and Sun proposed a generalized Gershgorin theorem for a regular matrix pair (see Corollary 2.5 in [113, Section 6.2.2]). We extend this famous theorem to the tensor case in this subsection. Define the p-norm of a tensor as the p-norm of its mode-1 unfolding [49, Chapter 12.4]

$$\|\mathcal{A}\|_p := \|\mathcal{A}_{(1)}\|_p.$$

Particularly, the tensor ∞-norm has the form

$$\|\mathcal{A}\|_\infty = \max_{i=1,2,\ldots,n} \sum_{i_2=1}^{n} \sum_{i_3=1}^{n} \cdots \sum_{i_m=1}^{n} |a_{ii_2\ldots i_m}|.$$

Notice that for a positive integer k, we have $\|\mathbf{x}^{[k]}\|_\infty = \|\mathbf{x}^{\otimes k}\|_\infty$, where $\mathbf{x}^{[k]} = [x_1^k, x_2^k, \ldots, x_n^k]^\top$ is the componentwise power and $\mathbf{x}^{\otimes k} = \mathbf{x} \otimes \mathbf{x} \otimes \cdots \otimes \mathbf{x}$ is the Kronecker product of k copies of $\mathbf{x}$ [49, Section 12.3]. Denote $M\mathcal{A} := \mathcal{A} \times_1 M^\top$ and $\mathcal{A}M^{m-1} = \mathcal{A} \times_2 M \cdots \times_m M$ for simplicity. Then we can prove the following lemma for ∞-norm:

Lemma 2.2. *Let $\{\mathcal{A}, \mathcal{B}\}$ and $\{\mathcal{C}, \mathcal{D}\} = \{C\mathcal{I}, D\mathcal{I}\}$ be two m^{th}-order n-dimensional regular tensor pairs, where C and D are two matrices. If $(\alpha, \beta) \in \mathbb{C}_{1,2}$ is an eigenvalue of $\{\mathcal{A}, \mathcal{B}\}$, then either $\det(\beta C - \alpha D) = 0$ or*

$$\left\|(\beta C - \alpha D)^{-1}[\beta(\mathcal{C} - \mathcal{A}) - \alpha(\mathcal{D} - \mathcal{B})]\right\|_\infty \geq 1.$$

Proof. If $(\alpha, \beta) \in \mathbb{C}_{1,2}$ satisfies that $\det(\beta\mathcal{C} - \alpha\mathcal{D}) \neq 0$, then it holds for an arbitrary nonzero vector $\mathbf{y} \in \mathbb{C}^n$ that $(\beta\mathcal{C} - \alpha\mathcal{D})\mathbf{y}^{m-1} \neq \mathbf{0}$, which is equivalent to $(\beta\mathcal{C} - \alpha\mathcal{D})\mathbf{y}^{[m-1]} \neq \mathbf{0}$. Then the matrix $\beta\mathcal{C} - \alpha\mathcal{D}$ is nonsingular.

Since (α, β) is an eigenvalue of $\{\mathcal{A}, \mathcal{B}\}$, there exists a nonzero vector $\mathbf{x} \in \mathbb{C}^n$ such that $\beta\mathcal{A}\mathbf{x}^{m-1} = \alpha\mathcal{B}\mathbf{x}^{m-1}$. This indicates that

$$\left[\beta(\mathcal{C} - \mathcal{A}) - \alpha(\mathcal{D} - \mathcal{B})\right]\mathbf{x}^{m-1} = (\beta\mathcal{C} - \alpha\mathcal{D})\mathbf{x}^{m-1}.$$

Thus from the expressions of $\mathcal{C}$ and $\mathcal{D}$, we have

$$\mathcal{G}\mathbf{x}^{m-1} := (\beta\mathcal{C} - \alpha\mathcal{D})^{-1}\left[\beta(\mathcal{C} - \mathcal{A}) - \alpha(\mathcal{D} - \mathcal{B})\right]\mathbf{x}^{m-1} = \mathbf{x}^{[m-1]}.$$

Then following the definition of the tensor ∞-norm and

$$\|\mathcal{G}\|_\infty = \max_{\mathbf{v} \in \mathbb{C}^{n^{m-1}}\setminus\{\mathbf{0}\}} \frac{\|\mathcal{G}_{(1)}\mathbf{v}\|_\infty}{\|\mathbf{v}\|_\infty} \geq \frac{\|\mathcal{G}_{(1)}\mathbf{x}^{\otimes(m-1)}\|_\infty}{\|\mathbf{x}^{\otimes(m-1)}\|_\infty} = \frac{\|\mathcal{G}\mathbf{x}^{m-1}\|_\infty}{\|\mathbf{x}^{[m-1]}\|_\infty} = 1,$$

we prove the result. $\qquad\square$

Comparing with Theorem 2.3 in [113, Section 6.2.2], Lemma 2.2 specifies that $\{\mathcal{C}, \mathcal{D}\}$ has a special structure and the norm in the result must be the ∞-norm. Nevertheless, these restrictions do not obstruct its application to the proof of the following lemma:

Lemma 2.3. *Let $\{\mathcal{A}, \mathcal{B}\}$ be an mth-order n-dimensional regular tensor pair. Furthermore, assume that $(a_{ii\ldots i}, b_{ii\ldots i}) \neq (0,0)$ for $i = 1, 2, \ldots, n$. Denote*

$$\mathscr{D}_i(\mathcal{A}, \mathcal{B}) := \left\{(\alpha, \beta) \in \mathbb{C}_{1,2} : |\beta a_{ii\ldots i} - \alpha b_{ii\ldots i}| \leq \sum_{\substack{(i_2,\ldots,i_m) \\ \neq (i,\ldots,i)}} |\beta a_{ii_2\ldots i_m} - \alpha b_{ii_2\ldots i_m}|\right\}$$

for $i = 1, 2, \ldots, n$. Then $\lambda(\mathcal{A}, \mathcal{B}) \subseteq \bigcup_{i=1}^n \mathscr{D}_i(\mathcal{A}, \mathcal{B})$.

Proof. Take the diagonal matrices $C = \text{diag}(a_{11\ldots 1}, a_{22\ldots 2}, \ldots, a_{nn\ldots n})$ and $D = \text{diag}(b_{11\ldots 1}, b_{22\ldots 2}, \ldots, b_{nn\ldots n})$ in Lemma 2.2. Then from the assumption, we know that $\{\mathcal{C}, \mathcal{D}\}$ is a regular tensor pair.

When $(\alpha, \beta) \in \lambda(\mathcal{A}, \mathcal{B})$ satisfies that $\det(\beta\mathcal{C} - \alpha\mathcal{D}) = 0$, it must hold that $\beta a_{ii\ldots i} - \alpha b_{ii\ldots i} = 0$ for some i. Thus it is obvious that $(\alpha, \beta) \in \mathscr{D}_i(\mathcal{A}, \mathcal{B})$.

When $(\alpha, \beta) \in \lambda(\mathcal{A}, \mathcal{B})$ does not satisfy that $\det(\beta\mathcal{C} - \alpha\mathcal{D}) = 0$, by Lemma 2.2, we have

$$\left\|(\beta\mathcal{C} - \alpha\mathcal{D})^{-1}\left[\beta(\mathcal{C} - \mathcal{A}) - \alpha(\mathcal{D} - \mathcal{B})\right]\right\|_\infty$$

$$= \max_{i=1,2,\ldots,n} \sum_{\substack{(i_2,\ldots,i_m) \\ \neq (i,\ldots,i)}} \left|\frac{\beta a_{ii_2\ldots i_m} - \alpha b_{ii_2\ldots i_m}}{\beta a_{ii\ldots i} - \alpha b_{ii\ldots i}}\right| \geq 1.$$

Thus for some i, it holds that

$$|\beta a_{ii\ldots i} - \alpha b_{ii\ldots i}| \leq \sum_{\substack{(i_2,\ldots,i_m) \\ \neq (i,\ldots,i)}} |\beta a_{ii_2\ldots i_m} - \alpha b_{ii_2\ldots i_m}|,$$

that is, $(\alpha, \beta) \in \mathscr{D}_i(\mathcal{A}, \mathcal{B})$. $\qquad\square$

Lemma 2.3 is alike in form to the Gershgorin circle theorem. However, it is not the desired result, since there are still α and β on the right-hand side of the inequality. To avoid this, we introduce the *chordal metric* in $\mathbb{C}_{1,2}$ [49, 113]:

$$\text{chord}\big((\alpha_1,\beta_1),(\alpha_2,\beta_2)\big) = \frac{|\alpha_1\beta_2 - \beta_1\alpha_2|}{\sqrt{|\alpha_1|^2+|\beta_1|^2}\sqrt{|\alpha_2|^2+|\beta_2|^2}}.$$

Moreover, we can easily prove by the Cauchy inequality that

$$\sum_{\substack{(i_2,i_3,\ldots,i_m)\\ \neq(i,i,\ldots,i)}} |\beta a_{ii_2\ldots i_m} - \alpha b_{ii_2\ldots i_m}| \leq \sqrt{|\alpha|^2+|\beta|^2}$$

$$\times \sqrt{\left(\sum_{\substack{(i_2,i_3,\ldots,i_m)\\ \neq(i,i,\ldots,i)}} |a_{ii_2\ldots i_m}|\right)^2 + \left(\sum_{\substack{(i_2,i_3,\ldots,i_m)\\ \neq(i,i,\ldots,i)}} |b_{ii_2\ldots i_m}|\right)^2}.$$

Finally, we obtain the Gershgorin circle theorem for regular tensor pairs.

Theorem 2.4. *Let $\{\mathcal{A},\mathcal{B}\}$ be an mth-order n-dimensional regular tensor pair. Suppose that $(a_{ii\ldots i}, b_{ii\ldots i}) \neq (0,0)$ for all $i=1,2,\ldots,n$. Denote the disks*

$$\mathscr{G}_i(\mathcal{A},\mathcal{B}) := \Big\{(\alpha,\beta)\in\mathbb{C}_{1,2} : \text{chord}\big((\alpha,\beta),(a_{ii\ldots i},b_{ii\ldots i})\big) \leq \gamma_i\Big\}$$

for $i=1,2,\ldots,n$, where

$$\gamma_i = \sqrt{\frac{\left(\sum_{\substack{(i_2,i_3,\ldots,i_m)\\ \neq(i,i,\ldots,i)}} |a_{ii_2\ldots i_m}|\right)^2 + \left(\sum_{\substack{(i_2,i_3,\ldots,i_m)\\ \neq(i,i,\ldots,i)}} |b_{ii_2\ldots i_m}|\right)^2}{|a_{ii\ldots i}|^2 + |b_{ii\ldots i}|^2}}.$$

Then $\lambda(\mathcal{A},\mathcal{B}) \subseteq \bigcup_{i=1}^n \mathscr{G}_i(\mathcal{A},\mathcal{B})$.

If $\mathcal{B}$ is taken as the unit tensor $\mathcal{I}$, then Theorem 2.4 reduces to the Gershgorin circle theorem for single tensors, that is, [98, Theorem 6(a)].

Furthermore, the Gershgorin circle theorem can have a tighter version when the order of the tensor pair is no less than 3. A tensor is called *semi-symmetric* [86] if its entries are invariable under any permutations of the last $(m-1)$ indices. Then for an arbitrary tensor $\mathcal{A}$, there is a semi-symmetric tensor $\tilde{\mathcal{A}}$ such that $\mathcal{A}x^{m-1} = \tilde{\mathcal{A}}x^{m-1}$ for all $x \in \mathbb{C}^n$. Concretely, the entries of $\tilde{\mathcal{A}}$ are

$$\tilde{a}_{ii_2\ldots i_m} = \frac{1}{|\pi(i_2,\ldots,i_m)|} \sum_{\substack{(i_2',i_3',\ldots,i_m')\in\\ \pi(i_2,i_3,\ldots,i_m)}} a_{ii_2'i_3'\ldots i_m'},$$

where $\pi(i_2,i_3,\ldots,i_m)$ denotes the set of all permutations of $(i_2,i_3,\ldots,i_m)$ and $|\pi(i_2,i_3,\ldots,i_m)|$ denotes the number of elements in $\pi(i_2,i_3,\ldots,i_m)$. Note that $a_{ii\ldots i} = \tilde{a}_{ii\ldots i}$ and

$$\sum_{\substack{(i_2,i_3,\ldots,i_m)\\ \neq(i,i,\ldots,i)}} |a_{ii_2\ldots i_m}| \geq \sum_{\substack{(i_2,i_3,\ldots,i_m)\\ \neq(i,i,\ldots,i)}} |\tilde{a}_{ii_2\ldots i_m}|.$$

Hence we have a tight version of the Gershgorin circle theorem for regular tensor pairs, that is, the disk $\widetilde{\mathscr{G}}_i(\mathcal{A},\mathcal{B})$ in the following theorem must be no larger than the disk $\mathscr{G}_i(\mathcal{A},\mathcal{B})$ in Theorem 2.4.

Theorem 2.5. *Let* $\{\mathcal{A}, \mathcal{B}\}$ *be an* m^{th}*-order* n*-dimensional regular tensor pair. Assume that* $(a_{ii\ldots i}, b_{ii\ldots i}) \neq (0, 0)$ *for all* $i = 1, 2, \ldots, n$. *Let* $\widetilde{A}$ *and* $\widetilde{B}$ *be the semi-symmetric tensors such that* $\mathcal{A}\mathbf{x}^{m-1} = \widetilde{A}\mathbf{x}^{m-1}$ *and* $\mathcal{B}\mathbf{x}^{m-1} = \widetilde{B}\mathbf{x}^{m-1}$ *for all* $\mathbf{x} \in \mathbb{C}^n$. *Denote the disks*

$$\widetilde{\mathscr{G}}_i(\mathcal{A}, \mathcal{B}) := \left\{ (\alpha, \beta) \in \mathbb{C}_{1,2} : \text{chord}\big((\alpha, \beta), (a_{ii\ldots i}, b_{ii\ldots i})\big) \leq \widetilde{\gamma}_i \right\}$$

for $i = 1, 2, \ldots, n$, *where*

$$\widetilde{\gamma}_i = \sqrt{\frac{\left(\sum_{\substack{(i_2, i_3, \ldots, i_m) \\ \neq (i,i,\ldots,i)}} |\widetilde{a}_{ii_2\ldots i_m}|\right)^2 + \left(\sum_{\substack{(i_2, i_3, \ldots, i_m) \\ \neq (i,i,\ldots,i)}} |\widetilde{b}_{ii_2\ldots i_m}|\right)^2}{|a_{ii\ldots i}|^2 + |b_{ii\ldots i}|^2}}.$$

Then $\lambda(\mathcal{A}, \mathcal{B}) \subseteq \bigcup_{i=1}^{n} \widetilde{\mathscr{G}}_i(\mathcal{A}, \mathcal{B})$.

2.3.5 Backward Error Analysis

Backward error is a measurement of the stability of a numerical algorithm [53]. It is also widely employed as a stopping criteria for iterative algorithms. We propose the framework of the backward error analysis (see [52] for the matrix version) on the algorithms for generalized tensor eigenproblems.

Suppose that we obtain a computational eigenvalue $(\widehat{\alpha}, \widehat{\beta})$ and the corresponding computed eigenvector $\widehat{\mathbf{x}}$ of a regular tensor pair $\{\mathcal{A}, \mathcal{B}\}$ by an algorithm. Then they can be regarded as an exact eigenvalue and eigenvector of another tensor pair $\{\mathcal{A} + \mathcal{E}, \mathcal{B} + \mathcal{F}\}$. Define the *normwise backward error* of the computed solution as

$$\eta_{\delta_1, \delta_2}(\widehat{\alpha}, \widehat{\beta}, \widehat{\mathbf{x}}) := \min\left\{ \|(\mathcal{E}/\delta_1, \mathcal{F}/\delta_2)\|_F : \widehat{\beta}(\mathcal{A} + \mathcal{E})\widehat{\mathbf{x}}^{m-1} = \widehat{\alpha}(\mathcal{B} + \mathcal{F})\widehat{\mathbf{x}}^{m-1} \right\}.$$

Here δ_1 and δ_2 are two parameters. When taking $\delta_1 = \delta_2 = 1$, the backward error is called the *absolute backward error* and denoted as $\eta_{[\text{abs}]}(\widehat{\alpha}, \widehat{\beta}, \widehat{\mathbf{x}})$. When taking $\delta_1 = \|\mathcal{A}\|_F$ and $\delta_2 = \|\mathcal{B}\|_F$, the backward error is called the *relative backward error* and denoted as $\eta_{[\text{rel}]}(\widehat{\alpha}, \widehat{\beta}, \widehat{\mathbf{x}})$.

Denote the residual $\widehat{\mathbf{r}} := \widehat{\alpha}\mathcal{B}\widehat{\mathbf{x}}^{m-1} - \widehat{\beta}\mathcal{A}\widehat{\mathbf{x}}^{m-1}$. Then the two tensors $\mathcal{E}$ and $\mathcal{F}$ must satisfy that $\widehat{\beta}\mathcal{E}\widehat{\mathbf{x}}^{m-1} - \widehat{\alpha}\mathcal{F}\widehat{\mathbf{x}}^{m-1} = \widehat{\mathbf{r}}$, which can be rewritten into an underdetermined linear equation [53, Chapter 21]

$$\left(\mathcal{E}_{(1)}/\delta_1, -\mathcal{F}_{(1)}/\delta_2\right) \cdot \begin{pmatrix} \delta_1 \widehat{\beta}\mathbf{x}^{\otimes(m-1)} \\ \delta_2 \widehat{\alpha}\mathbf{x}^{\otimes(m-1)} \end{pmatrix} = \widehat{\mathbf{r}}.$$

Hence the least-norm solution is given by

$$\left(\mathcal{E}_{(1)}/\delta_1, -\mathcal{F}_{(1)}/\delta_2\right) = \widehat{\mathbf{r}} \begin{pmatrix} \delta_1 \widehat{\beta}\mathbf{x}^{\otimes(m-1)} \\ \delta_2 \widehat{\alpha}\mathbf{x}^{\otimes(m-1)} \end{pmatrix}^{\dagger},$$

where $A^{\dagger}$ denotes the Moore-Penrose pseudoinverse of a matrix A [49, Section 5.5.2]. Notice that $\mathbf{v}^{\dagger} = \mathbf{v}^*/\|\mathbf{v}\|_2^2$ and $\|\widehat{\mathbf{x}}^{\otimes(m-1)}\|_2 = \|\widehat{\mathbf{x}}\|_2^{m-1}$. Therefore we derive the explicit form of the normwise backward error.

Theorem 2.6. *The normwise backward error is given by*

$$(2.2) \quad \eta_{\delta_1,\delta_2}(\widehat{\alpha},\widehat{\beta},\widehat{\mathbf{x}}) = \left\| \left(\mathcal{E}_{(1)}/\delta_1, -\mathcal{F}_{(1)}/\delta_2 \right) \right\|_F = \frac{1}{\sqrt{|\delta_1\widehat{\beta}|^2 + |\delta_2\widehat{\alpha}|^2}} \cdot \frac{\|\widehat{\mathbf{r}}\|_2}{\|\widehat{\mathbf{x}}\|_2^{m-1}}.$$

Particularly, the absolute backward error is presented by

$$(2.3) \quad \eta_{[\text{abs}]}(\widehat{\alpha},\widehat{\beta},\widehat{\mathbf{x}}) = \frac{1}{\sqrt{|\widehat{\beta}|^2 + |\widehat{\alpha}|^2}} \cdot \frac{\|\widehat{\mathbf{r}}\|_2}{\|\widehat{\mathbf{x}}\|_2^{m-1}},$$

and the relative backward error can be expressed as

$$(2.4) \quad \eta_{[\text{rel}]}(\widehat{\alpha},\widehat{\beta},\widehat{\mathbf{x}}) = \frac{1}{\sqrt{|\widehat{\beta}|^2\|\mathcal{A}\|_F^2 + |\widehat{\alpha}|^2\|\mathcal{B}\|_F^2}} \cdot \frac{\|\widehat{\mathbf{r}}\|_2}{\|\widehat{\mathbf{x}}\|_2^{m-1}}.$$

Furthermore, we can also discuss the componentwise backward error (see [52] for the matrix version). Define the *componentwise backward error* corresponding to the nonnegative weight tensors $\mathcal{U}, \mathcal{V}$ of the computed solution as

$$\omega_{\mathcal{U},\mathcal{V}}(\widehat{\alpha},\widehat{\beta},\widehat{\mathbf{x}}) := \min\left\{ \epsilon : \widehat{\beta}(\mathcal{A}+\mathcal{E})\widehat{\mathbf{x}}^{m-1} = \widehat{\alpha}(\mathcal{B}+\mathcal{F})\widehat{\mathbf{x}}^{m-1}, |\mathcal{E}| \leq \epsilon\mathcal{U}, |\mathcal{F}| \leq \epsilon\mathcal{V} \right\}.$$

Take the absolute values on both sides of $\widehat{\mathbf{r}} = \widehat{\beta}\mathcal{E}\widehat{\mathbf{x}}^{m-1} - \widehat{\alpha}\mathcal{F}\widehat{\mathbf{x}}^{m-1}$ and obtain

$$|\widehat{\mathbf{r}}| \leq |\widehat{\beta}||\mathcal{E}||\widehat{\mathbf{x}}|^{m-1} + |\widehat{\alpha}||\mathcal{F}||\widehat{\mathbf{x}}|^{m-1} \leq \epsilon\left(|\widehat{\beta}|\mathcal{U}|\widehat{\mathbf{x}}|^{m-1} + |\widehat{\alpha}|\mathcal{V}|\widehat{\mathbf{x}}|^{m-1} \right).$$

Hence the componentwise backward error has a lower bound

$$\omega_{\mathcal{U},\mathcal{V}}(\widehat{\alpha},\widehat{\beta},\widehat{\mathbf{x}}) \geq \max_{i=1,2,\dots,n} \frac{|\widehat{r}_i|}{\left(|\widehat{\beta}|\mathcal{U}|\widehat{\mathbf{x}}|^{m-1} + |\widehat{\alpha}|\mathcal{V}|\widehat{\mathbf{x}}|^{m-1} \right)_i}.$$

The equality can be attained by

$$\mathcal{E} = \epsilon \cdot \text{sign}(\widehat{\beta})^{-1} \cdot SUT^{m-1} \quad \text{and} \quad \mathcal{F} = \epsilon \cdot -\text{sign}(\widehat{\alpha})^{-1} \cdot SVT^{m-1},$$

where S and T are two diagonal matrices such that $\widehat{\mathbf{r}} = S|\widehat{\mathbf{r}}|$ and $T\widehat{\mathbf{x}} = |\widehat{\mathbf{x}}|$. Therefore we derive the explicit form of the componentwise backward error.

Theorem 2.7. *The componentwise backward error is given by*

$$(2.5) \quad \omega_{\mathcal{U},\mathcal{V}}(\widehat{\alpha},\widehat{\beta},\widehat{\mathbf{x}}) = \max_{i=1,2,\dots,n} \frac{|\widehat{r}_i|}{\left(|\widehat{\beta}|\mathcal{U}|\widehat{\mathbf{x}}|^{m-1} + |\widehat{\alpha}|\mathcal{V}|\widehat{\mathbf{x}}|^{m-1} \right)_i}.$$

2.4   Real Tensor Pairs

For tensor problems, the real case is quite different from the complex one. For instance, an eigenvector associated with a real eigenvalue of a real tensor can be complex and cannot be transformed into a real one by multiplying a scalar, which will not occur for matrices. Therefore we need to discuss the real tensor pairs separately.

2.4.1 The Crawford Number

We need to define the concept of "regular pair" in another way for real tensor pairs. Let $\mathcal{A}$ and $\mathcal{B}$ be two m^{th}-order real tensors in $\mathbb{R}^{n \times n \times \cdots \times n}$. If there does not exist a real vector $\mathbf{x} \in \mathbb{R}^n \setminus \{\mathbf{0}\}$ such that

$$\mathcal{A}\mathbf{x}^{m-1} = \mathcal{B}\mathbf{x}^{m-1} = \mathbf{0},$$

then we call $\{\mathcal{A}, \mathcal{B}\}$ an *R-regular tensor pair* ("R" for "real"). A real eigenvalue of an R-regular tensor pair with a real eigenvector is called an *H-eigenvalue*, following the same name for single tensors [98]. Moreover, Zhang [134] proved the existence of an H-eigenvalue for a real tensor pair, when m is even and n is odd.

Obviously, a tensor pair is R-regular if it is regular. However, an R-regular tensor pair need not be regular.

Example 2.2. Suppose that m is even. Let $\mathcal{Z}$ be a real tensor with

$$z_{i_1 i_2 \ldots i_m} = \delta_{i_1 i_2} \delta_{i_3 i_4} \cdots \delta_{i_{m-1} i_m}, \qquad \delta_{ij} = \begin{cases} 1, & i = j, \\ 0, & i \neq j. \end{cases}$$

Then $\mathcal{Z}\mathbf{x}^{m-1} = \mathbf{x} \cdot (\mathbf{x}^\top \mathbf{x})^{(m-2)/2}$. As stated in [20], the H-eigenvalues of $\{\mathcal{A}, \mathcal{Z}\}$ are the Z-eigenvalues of $\mathcal{A}$. For an arbitrary $\mathbf{x} \in \mathbb{R}^n \setminus \{\mathbf{0}\}$, it holds that $\mathcal{Z}\mathbf{x}^{m-1} \neq \mathbf{0}$ since $\mathbf{x}^\top \mathbf{x} > 0$. Thus the real tensor pair $\{\mathcal{A}, \mathcal{Z}\}$ is R-regular. Nevertheless, $\mathbf{y}^\top \mathbf{y} = 0$ can be true for some complex vector $\mathbf{y} \in \mathbb{C}^n \setminus \{\mathbf{0}\}$. Thus if $\mathcal{A}\mathbf{y}^{m-1} = \mathbf{0}$, then $\{\mathcal{A}, \mathcal{Z}\}$ is a singular tensor pair.

Define the Crawford number (see [113, Section 6.1.3] for the matrix case) of a real tensor pair $\{\mathcal{A}, \mathcal{B}\}$ as

$$c_r(\mathcal{A}, \mathcal{B}) := \min \left\{ \sqrt{(\mathcal{A}\mathbf{x}^m)^2 + (\mathcal{B}\mathbf{x}^m)^2} : \|\mathbf{x}\| = 1, \mathbf{x} \in \mathbb{R}^n \right\}.$$

If $c_r(\mathcal{A}, \mathcal{B}) > 0$, then we call $\{\mathcal{A}, \mathcal{B}\}$ a *definite tensor pair*. A definite tensor pair must be an R-regular pair, since $(\mathcal{A}\mathbf{x}^m)^2 + (\mathcal{B}\mathbf{x}^m)^2 = 0$ if $\mathcal{A}\mathbf{x}^{m-1} = \mathcal{B}\mathbf{x}^{m-1} = \mathbf{0}$.

We will now discuss the continuity of the Crawford number. Denote

$$\|\mathcal{A}\|_L := \max \left\{ \left| \mathcal{A} \times_1 \mathbf{x}_1 \times_2 \mathbf{x}_2 \cdots \times_m \mathbf{x}_m \right| : \|\mathbf{x}_k\| = 1, k = 1, 2, \ldots, m \right\},$$

which is a norm on $\mathbb{R}^{n \times n \times \cdots \times n}$ defined by Lim [76]. Here, the norm in the definition can be an arbitrary vector norm if not specified. Then the continuity of the Crawford number can be seen from the following theorem.

Theorem 2.8. *Let $\{\mathcal{A}, \mathcal{B}\}$ and $\{\mathcal{A} + \mathcal{E}, \mathcal{B} + \mathcal{F}\}$ be two real tensor pairs. Then*

$$(2.6) \qquad \left| c_r(\mathcal{A} + \mathcal{E}, \mathcal{B} + \mathcal{F}) - c_r(\mathcal{A}, \mathcal{B}) \right| \leq \sqrt{\|\mathcal{E}\|_L^2 + \|\mathcal{F}\|_L^2}.$$

Proof. By the definition of the Crawford number, we have

$$c_r(\mathcal{A} + \mathcal{E}, \mathcal{B} + \mathcal{F}) = \min_{\|\mathbf{x}\|=1} \sqrt{\left((\mathcal{A} + \mathcal{E})\mathbf{x}^m\right)^2 + \left((\mathcal{B} + \mathcal{F})\mathbf{x}^m\right)^2}$$

$$\geq \min_{\|\mathbf{x}\|=1} \left[\sqrt{(\mathcal{A}\mathbf{x}^m)^2 + (\mathcal{B}\mathbf{x}^m)^2} - \sqrt{(\mathcal{E}\mathbf{x}^m)^2 + (\mathcal{F}\mathbf{x}^m)^2} \right]$$

$$\geq c_r(\mathcal{A}, \mathcal{B}) - \sqrt{\max_{\|\mathbf{x}\|=1} (\mathcal{E}\mathbf{x}^m)^2 + \max_{\|\mathbf{x}\|=1} (\mathcal{F}\mathbf{x}^m)^2}.$$

$$= c_r(\mathcal{A}, \mathcal{B}) - \sqrt{\|\mathcal{E}\|_L^2 + \|\mathcal{F}\|_L^2}$$

The remaining part follows directly by

$$
\begin{aligned}
c_r(\mathcal{A}, \mathcal{B}) &= c_r(\mathcal{A} + \mathcal{E} - \mathcal{E}, \mathcal{B} + \mathcal{F} - \mathcal{F}) \\
&\geq c_r(\mathcal{A} + \mathcal{E}, \mathcal{B} + \mathcal{F}) - \sqrt{\max_{\|\mathbf{x}\|=1}(-\mathcal{E}\mathbf{x}^m)^2 + \max_{\|\mathbf{x}\|=1}(-\mathcal{F}\mathbf{x}^m)^2} \\
&= c_r(\mathcal{A} + \mathcal{E}, \mathcal{B} + \mathcal{F}) - \sqrt{\|\mathcal{E}\|_L^2 + \|\mathcal{F}\|_L^2}.
\end{aligned}
$$

$\square$

By the same approach, we prove that small perturbations preserve the definiteness of a real tensor pair.

Theorem 2.9. *Let* $\{\mathcal{A}, \mathcal{B}\}$ *be a definite tensor pair and* $\{\mathcal{A} + \mathcal{E}, \mathcal{B} + \mathcal{F}\}$ *be a real tensor pair. If*

$$
\max_{\|\mathbf{x}\|=1} \sqrt{\frac{(\mathcal{E}\mathbf{x}^m)^2 + (\mathcal{F}\mathbf{x}^m)^2}{(\mathcal{A}\mathbf{x}^m)^2 + (\mathcal{B}\mathbf{x}^m)^2}} < 1,
$$

then $\{\mathcal{A} + \mathcal{E}, \mathcal{B} + \mathcal{F}\}$ *is also a definite tensor pair.*

Proof. By the definition of the Crawford number, we have

$$
\begin{aligned}
c_r(\mathcal{A} + \mathcal{E}, \mathcal{B} + \mathcal{F}) &= \min_{\|\mathbf{x}\|=1} \sqrt{\left((\mathcal{A} + \mathcal{E})\mathbf{x}^m\right)^2 + \left((\mathcal{B} + \mathcal{F})\mathbf{x}^m\right)^2} \\
&\geq \min_{\|\mathbf{x}\|=1} \left[\sqrt{(\mathcal{A}\mathbf{x}^m)^2 + (\mathcal{B}\mathbf{x}^m)^2} - \sqrt{(\mathcal{E}\mathbf{x}^m)^2 + (\mathcal{F}\mathbf{x}^m)^2} \right] \\
&\geq c_r(\mathcal{A}, \mathcal{B}) \left(1 - \max_{\|\mathbf{x}\|=1} \sqrt{\frac{(\mathcal{E}\mathbf{x}^m)^2 + (\mathcal{F}\mathbf{x}^m)^2}{(\mathcal{A}\mathbf{x}^m)^2 + (\mathcal{B}\mathbf{x}^m)^2}} \right).
\end{aligned}
$$

When the item in the blanket is positive, we can conclude that $c_r(\mathcal{A}+\mathcal{E}, \mathcal{B}+\mathcal{F}) > 0$, since $c_r(\mathcal{A}, \mathcal{B}) > 0$. $\square$

A more concise and useful corollary also follows directly.

Corollary 2.10. *Let* $\{\mathcal{A}, \mathcal{B}\}$ *be a definite tensor pair and* $\{\mathcal{A} + \mathcal{E}, \mathcal{B} + \mathcal{F}\}$ *be a real tensor pair. If*

$$
(2.7) \qquad\qquad \sqrt{\|\mathcal{E}\|_L^2 + \|\mathcal{F}\|_L^2} < c_r(\mathcal{A}, \mathcal{B}),
$$

then $\{\mathcal{A} + \mathcal{E}, \mathcal{B} + \mathcal{F}\}$ *is also a definite tensor pair.*

2.4.2   Symmetric-Definite Tensor Pairs

A tensor $\mathcal{A}$ is called *symmetric* [98] if its entries are invariable under any permutations of the indices and is called *positive definite* [98] if $\mathcal{A}\mathbf{x}^m > 0$ for all $\mathbf{x} \in \mathbb{R}^n \setminus \{\mathbf{0}\}$. (Thus m must be even.) Then we call $\{\mathcal{A}, \mathcal{B}\}$ a *symmetric-definite tensor pair* if $\mathcal{A}$ is symmetric and $\mathcal{B}$ is symmetric positive definite (see [49, Section 8.7.1] for the matrix case). It is easy to understand that a symmetric-definite tensor pair must be a definite pair as introduced in Section 2.4.1. Chang et al. [20] proved the existence of the H-eigenvalues for symmetric-definite tensor pairs. Furthermore, there is a clear depiction of the maximal and the minimal H-eigenvalues of a symmetric-definite tensor pair. Before stating the main result, we need some lemmata.

Lemma 2.11. *Let $\mathcal{S}$ and $\mathcal{T}$ be two m^{th}-order n-dimensional real symmetric tensors. Assume that m is even and $\mathcal{T}$ is positive definite. Define*

$$W_{\mathcal{T}}(\mathcal{S}) := \left\{ \mathcal{S}\mathbf{x}^m : \mathcal{T}\mathbf{x}^m = 1, \mathbf{x} \in \mathbb{R}^n \right\}.$$

Then $W_{\mathcal{T}}(\mathcal{S})$ is a convex set.

Proof. We prove this lemma following the definition of convex sets. Let $\mathbf{x}_1, \mathbf{x}_2 \in \mathbb{R}^n$ satisfy that $\mathcal{T}\mathbf{x}_1^m = \mathcal{T}\mathbf{x}_2^m = 1$. Then $\omega_1 = \mathcal{S}\mathbf{x}_1^m$ and $\omega_2 = \mathcal{S}\mathbf{x}_2^m$ are two points in $W_{\mathcal{T}}(\mathcal{S})$. Without loss of generality, we can assume that $\omega_1 > \omega_2$. For an arbitrary $t \in (0, 1)$, we desire a vector $\widetilde{\mathbf{x}}$ with $\mathcal{S}\widetilde{\mathbf{x}}^m = t\omega_1 + (1-t)\omega_2$ and $\mathcal{T}\widetilde{\mathbf{x}}^m = 1$, so that $t\omega_1 + (1-t)\omega_2$ is also a point in $W_{\mathcal{T}}(\mathcal{S})$.

Take $\mathbf{x} = \tau\mathbf{x}_1 + \mathbf{x}_2$, and τ is selected to satisfy that

$$\mathcal{S}\mathbf{x}^m = [t\omega_1 + (1-t)\omega_2] \cdot \mathcal{T}\mathbf{x}^m.$$

Recall that $\mathcal{A}(\mathbf{u} + \mathbf{v})^m$ can be expanded for a symmetric tensor $\mathcal{A}$

$$\mathcal{A}(\mathbf{u} + \mathbf{v})^m = \sum_{k=0}^{m} \binom{m}{k} \mathcal{A}\mathbf{u}^k\mathbf{v}^{m-k},$$

where $\mathcal{A}\mathbf{u}^k\mathbf{v}^{m-k} = \mathcal{A} \times_1 \mathbf{u} \cdots \times_k \mathbf{u} \times_{k+1} \mathbf{v} \cdots \times_m \mathbf{v}$.

Thus $p(\tau) = \mathcal{S}\mathbf{x}^m - [t\omega_1 + (1-t)\omega_2] \cdot \mathcal{T}\mathbf{x}^m$ is a polynomial about τ of degree m. Moreover, the coefficient of the item τ^m is

$$\mathcal{S}\mathbf{x}_1^m - [t\omega_1 + (1-t)\omega_2] \cdot \mathcal{T}\mathbf{x}_1^m = (1-t)(\omega_1 - \omega_2) > 0,$$

and the coefficient of the item 1 is

$$\mathcal{S}\mathbf{x}_2^m - [t\omega_1 + (1-t)\omega_2] \cdot \mathcal{T}\mathbf{x}_2^m = t(\omega_2 - \omega_1) < 0.$$

Therefore there must be a real τ such that $p(\tau) = 0$. With such a τ, we can take $\widetilde{\mathbf{x}} = \mathbf{x}/(\mathcal{T}\mathbf{x}^m)^{1/m}$, since $\mathcal{T}$ is positive definite. Hence we prove this lemma. $\square$

When $\mathcal{T}$ is taken as $\mathcal{I}$, we know from Lemma 2.11 that $W_{\mathcal{I}}(\mathcal{S})$ is convex. Furthermore, $W_{\mathcal{I}}(\mathcal{S})$ is also compact, since the set $\{\mathbf{x} \in \mathbb{R}^n : \mathcal{I}\mathbf{x}^m = 1\}$, that is, $\{\mathbf{x} \in \mathbb{R}^n : \|\mathbf{x}\|_m = 1\}$, is compact and the function $\mathbf{x} \mapsto \mathcal{S}\mathbf{x}^m$ is continuous. Thus the set $W_{\mathcal{I}}(\mathcal{S})$ is actually a segment on the real axis.

Remark 2.12. *For a real matrix S, the set $\{\mathbf{x}^\top S\mathbf{x} : \mathbf{x}^\top \mathbf{x} = 1, \mathbf{x} \in \mathbb{R}^n\}$ is called the real field of values [55, Section 1.8.7]. Similarly, we define the real field of values for an even-order real tensor $\mathcal{S}$ as $W_{\mathcal{I}}(\mathcal{S})$. The above discussion reveals that $W_{\mathcal{I}}(\mathcal{S})$ is a segment on the real axis.*

Denote by

$$\omega_{\max} := \max \{\omega : \omega \in W_{\mathcal{I}}(\mathcal{S})\} \quad \text{and} \quad \omega_{\min} := \min \{\omega : \omega \in W_{\mathcal{I}}(\mathcal{S})\}.$$

If $\mathcal{S}$ is positive definite, then $\omega_{\max} \geq \omega_{\min} > 0$. Thus for any vector $\mathbf{x} \in \mathbb{R}^n$, we have

$$\omega_{\min}\mathcal{I}\mathbf{x}^m \leq \mathcal{S}\mathbf{x}^m \leq \omega_{\max}\mathcal{I}\mathbf{x}^m,$$

or, equivalently,

$$\left(\omega_{\max}^{-1}\mathcal{S}\mathbf{x}^m\right)^{1/m} \leq \|\mathbf{x}\|_m \leq \left(\omega_{\min}^{-1}\mathcal{S}\mathbf{x}^m\right)^{1/m},$$

which indicates that the set $\{\mathbf{x} \in \mathbb{R}^n : \mathcal{S}\mathbf{x}^m = 1\}$ is bounded. Therefore we prove that the set is compact if $\mathcal{S}$ is symmetric positive definite.

Consider the two optimization problems

$$
\text{(P1)} \quad
\begin{aligned}
\max \quad & \mathcal{A}\mathbf{x}^m, \\
\text{s.t.} \quad & \mathcal{B}\mathbf{x}^m = 1,
\end{aligned}
\qquad
\text{(P2)} \quad
\begin{aligned}
\min \quad & \mathcal{A}\mathbf{x}^m, \\
\text{s.t.} \quad & \mathcal{B}\mathbf{x}^m = 1.
\end{aligned}
$$

From the discussion above, we know that the feasible set $\{\mathbf{x} \in \mathbb{R}^n : \mathcal{B}\mathbf{x}^m = 1\}$ of (P1) and (P2) is compact, since $\mathcal{B}$ is positive definite. Moreover, the objective function $\mathbf{x} \mapsto \mathcal{A}\mathbf{x}^m$ is continuous. Therefore the maximum and the minimum of $\mathcal{A}\mathbf{x}^m$ under the restriction $\mathcal{B}\mathbf{x}^m = 1$ are attainable, and the maximizer and the minimizer must be KKT points [87]. Since $\mathcal{A}$ and $\mathcal{B}$ are symmetric, the KKT points of (P1) and (P2) satisfy that

$$
\nabla\left(\mathcal{A}\mathbf{x}^m - \lambda\mathcal{B}\mathbf{x}^m\right) = 0 \Leftrightarrow \mathcal{A}\mathbf{x}^{m-1} = \lambda\mathcal{B}\mathbf{x}^{m-1},
$$

which are exactly the eigenvectors of $\{\mathcal{A}, \mathcal{B}\}$ corresponding to the eigenvalues $\lambda = \mathcal{A}\mathbf{x}^m/\mathcal{B}\mathbf{x}^m$. Hence we have the following theorem as a generalization of the Rayleigh-Ritz theorem for a symmetric matrix [49, Section 8.1.1]. When we take $\mathcal{B} = \mathcal{I}$, this theorem reduces to the single tensor case, that is, [98, Theorem 5].

Theorem 2.13. *Let $\{\mathcal{A}, \mathcal{B}\}$ be an mth-order n-dimensional symmetric-definite tensor pair. Denote the maximal H-eigenvalue of $\{\mathcal{A}, \mathcal{B}\}$ as*

$$
\lambda_{\max} = \max\left\{\lambda \in \mathbb{R} : \mathcal{A}\mathbf{x}^{m-1} = \lambda\mathcal{B}\mathbf{x}^{m-1}, \mathbf{x} \in \mathbb{R}^n \setminus \{\mathbf{0}\}\right\},
$$

and the minimal H-eigenvalue of $\{\mathcal{A}, \mathcal{B}\}$ as

$$
\lambda_{\min} = \min\left\{\lambda \in \mathbb{R} : \mathcal{A}\mathbf{x}^{m-1} = \lambda\mathcal{B}\mathbf{x}^{m-1}, \mathbf{x} \in \mathbb{R}^n \setminus \{\mathbf{0}\}\right\}.
$$

Then we have

$$
(2.8) \qquad \lambda_{\max} = \max_{\mathbf{x} \in \mathbb{R}^n \setminus \{\mathbf{0}\}} \frac{\mathcal{A}\mathbf{x}^m}{\mathcal{B}\mathbf{x}^m}, \qquad \lambda_{\min} = \min_{\mathbf{x} \in \mathbb{R}^n \setminus \{\mathbf{0}\}} \frac{\mathcal{A}\mathbf{x}^m}{\mathcal{B}\mathbf{x}^m}.
$$

After obtaining these variational forms, we can discuss the perturbations of the maximal and the minimal H-eigenvalues of a symmetric-definite tensor pair. Since the proof is similar to that of Theorem 3.2 in [113, Section 6.3.1], we state the result directly. We write $\operatorname{chord}(\lambda, \mu)$ instead of $\operatorname{chord}\left((\lambda, 1), (\mu, 1)\right)$ for simplicity. We shall conclude from the following theorem that $1/c_r(\mathcal{A}, \mathcal{B})$ can be regarded as a condition number of the real generalized tensor eigenvalue problem.

Theorem 2.14. *Let $\{\mathcal{A}, \mathcal{B}\}$ and $\{\mathcal{A}+\mathcal{E}, \mathcal{B}+\mathcal{F}\}$ be two symmetric-definite tensor pairs. Denote the maximal and the minimal H-eigenvalues of $\{\mathcal{A}, \mathcal{B}\}$ as $\lambda_{\max}$ and $\lambda_{\min}$, respectively, and the ones for $\{\mathcal{A} + \mathcal{E}, \mathcal{B} + \mathcal{F}\}$ as $\widetilde{\lambda}_{\max}$ and $\widetilde{\lambda}_{\min}$. If the perturbation tensors $\mathcal{E}$ and $\mathcal{F}$ satisfy that*

$$
\sqrt{\|\mathcal{E}\|_L^2 + \|\mathcal{F}\|_L^2} < c_r(\mathcal{A}, \mathcal{B}),
$$

then the perturbations of the maximal and the minimal H-eigenvalues are bounded by

$$
\operatorname{chord}(\lambda_{\max}, \widetilde{\lambda}_{\max}) \leq \frac{\sqrt{\|\mathcal{E}\|_L^2 + \|\mathcal{F}\|_L^2}}{c_r(\mathcal{A}, \mathcal{B})}
$$

and

$$\text{chord}(\lambda_{\min}, \widetilde{\lambda}_{\min}) \leq \frac{\sqrt{\|\mathcal{E}\|_L^2 + \|\mathcal{F}\|_L^2}}{c_r(\mathcal{A}, \mathcal{B})}.$$

When the right-hand side tensor $\mathcal{B}$ is fixed, that is, there is no perturbation on $\mathcal{B}$, the next corollary extends the Weyl theorem [113, Section 4.4.2] and follows directly from Theorem 2.13. This result is also valid for H- and Z-eigenvalues for single tensors, taking $\mathcal{B}$ as $\mathcal{I}$ and the tensor $\mathcal{Z}$ in Example 2.2, respectively.

Corollary 2.15. *Let $\mathcal{A}$ and $\mathcal{E}$ be two m^{th}-order n-dimensional symmetric tensors. Then*

$$\lambda_{\max}(\mathcal{A}) + \lambda_{\min}(\mathcal{E}) \leq \lambda_{\max}(\mathcal{A} + \mathcal{E}) \leq \lambda_{\max}(\mathcal{A}) + \lambda_{\max}(\mathcal{E}),$$
$$\lambda_{\min}(\mathcal{A}) + \lambda_{\min}(\mathcal{E}) \leq \lambda_{\min}(\mathcal{A} + \mathcal{E}) \leq \lambda_{\min}(\mathcal{A}) + \lambda_{\max}(\mathcal{E}),$$

where $\lambda_{\max}$ and $\lambda_{\min}$ can stand for the extremal H- or Z-eigenvalues of a tensor.

Example 2.3. As stated in Example 2.2, the Z-eigenvalues of an even order tensor $\mathcal{A}$ are also the eigenvalues of the tensor pair $\{\mathcal{A}, \mathcal{Z}\}$. By Theorem 2.13, we have

$$\lambda_{\max} = \max_{\|\mathbf{x}\|_2=1, \mathbf{x} \in \mathbb{R}^n} \mathcal{A}\mathbf{x}^m, \quad \lambda_{\min} = \min_{\|\mathbf{x}\|_2=1, \mathbf{x} \in \mathbb{R}^n} \mathcal{A}\mathbf{x}^m,$$

which is a known result in [98]. Furthermore, because $\mathcal{Z}\mathbf{x}^m = 1$ for all $\mathbf{x}$ such that $\mathbf{x}^\top\mathbf{x} = 1$, the Crawford number of $\{\mathcal{A}, \mathcal{Z}\}$:

$$c_r(\mathcal{A}, \mathcal{Z}) = \sqrt{\min_{\|\mathbf{x}\|_2=1} (\mathcal{A}\mathbf{x}^m)^2 + 1} \geq 1.$$

Thus $\{\mathcal{A}, \mathcal{Z}\}$ must be a definite pair. Since $\mathcal{Z}$ is a fixed tensor, we assume that there is no perturbation on it. Therefore Theorem 2.14 tells us that if the perturbation $\mathcal{E}$ of $\mathcal{A}$ satisfies $\|\mathcal{E}\|_L \leq \epsilon$, then the perturbations of the maximal and the minimal Z-eigenvalues of $\mathcal{A}$ are bounded by

$$\text{chord}(\lambda_{\max}, \widetilde{\lambda}_{\max}) \leq \epsilon, \quad \text{chord}(\lambda_{\min}, \widetilde{\lambda}_{\min}) \leq \epsilon.$$

These perturbation bounds ensure that the extremal Z-eigenvalues may be computed very accurately by numerical methods in [30, 68].

Since there is no odd-order positive definite tensor, all the above discussions are about even-order tensors. Nevertheless, the applications with Z- or US-eigenvalues associate not only with even-order tensors. Then how do we treat the odd-order case? The following remark is devoted to this problem.

Remark 2.16. *We shall introduce the concept of abstract tensor to deal with the odd-order situation. Take the tensor pair approach for Z-eigenvalues as an example. In Examples 2.2 and 2.3, we define an even-order tensor $\mathcal{Z}$ such that*

$$(2.9) \qquad \mathcal{Z}\mathbf{x}^{m-1} = \mathbf{x} \cdot (\mathbf{x}^\top\mathbf{x})^{(m-2)/2}$$

for all real vector $\mathbf{x}$. Apparently, there is no odd order tensor satisfying (2.9), since $(m-2)/2$ is a fractional number. However, when we regard $\mathcal{Z}$ as a homogenous operator on $\mathbf{x}$, an mth-order tensor $\mathcal{Z}$ satisfying (2.9) can also be *defined implicitly for an odd m. Note that $\mathcal{Z}$ here is an operator rather than a concrete tensor, thus we call it an* abstract *tensor. Furthermore, if we define*

$$\mathcal{Z}\mathbf{x}^m := \mathbf{x}^\top (\mathcal{Z}\mathbf{x}^{m-1}) = (\mathbf{x}^\top \mathbf{x})^{m/2},$$

then the abstract tensor $\mathcal{Z}$ is also positive definite, and it also holds that $\nabla(\mathcal{Z}\mathbf{x}^m) = m\mathcal{Z}\mathbf{x}^{m-1}$. (This property is called weak symmetry in [20].) After a little analysis, we find that all the results in Section 2.4 are valid for this odd order abstract tensor pair $\{\mathcal{A}, \mathcal{Z}\}$. Therefore the properties of Z-eigenvalues discussed in Example 2.3 also hold for odd order tensors.

2.5   Sign-Complex Spectral Radius

The sign-complex spectral radius of a complex matrix is introduced by Rump [108] for extending the Perron-Frobenius theorem from nonnegative matrices to general complex matrices. We further extend this concept to complex tensor pairs in this section.

2.5.1   Definitions

Similarly to the θ-spectral radius in Section 2.3.2, we define the *sign-complex θ-spectral radius* of a regular tensor pair $\{\mathcal{A}, \mathcal{B}\}$ as

$$(2.10) \qquad \rho_\theta^{\mathbb{C}}(\mathcal{A}, \mathcal{B}) := \max \big\{ \theta(|\alpha|, |\beta|) : \big|\beta\mathcal{A}\mathbf{x}^{m-1}\big| = \big|\alpha\mathcal{B}\mathbf{x}^{m-1}\big|,$$
$$(\alpha, \beta) \in \mathbb{C}_{1,2}, \mathbf{x} \in \mathbb{C}^n \setminus \{\mathbf{0}\}\big\}.$$

Recall the relationship $\rho_\theta(\mathcal{A}, \mathcal{B}) \leq \psi_\theta(\mathcal{A}, \mathcal{B})$ in Section 2.3.2. Following the definition, we can easily verify that the sign-complex θ-spectral radius lies between the θ-spectral radius and the function $\psi_\theta(\mathcal{A}, \mathcal{B})$, that is,

$$\rho_\theta(\mathcal{A}, \mathcal{B}) \leq \rho_\theta^{\mathbb{C}}(\mathcal{A}, \mathcal{B}) \leq \psi_\theta(\mathcal{A}, \mathcal{B}).$$

A complex signature matrix refers to a unitary diagonal matrix, that is, $|S| = I$. Rump [108] pointed out that for any vector $\mathbf{x} \in \mathbb{C}^n$, there is a signature matrix S such that $S\mathbf{x} = |\mathbf{x}|$. If all the entries of the vector are nonzero, then this signature matrix is unique.

Thus there are two signature matrices S_1 and S_2 such that

$$S_1(\beta\mathcal{A}\mathbf{x}^{m-1}) = \big|\beta\mathcal{A}\mathbf{x}^{m-1}\big|, \quad S_2(\alpha\mathcal{B}\mathbf{x}^{m-1}) = \big|\alpha\mathcal{B}\mathbf{x}^{m-1}\big|.$$

Let $S = S_2^* S_1$, where S_2^* is the conjugate transpose of S_2. Then S is also a signature matrix, and we obtain the following equivalent definition of the sign-complex θ-spectral radius

$$(2.11) \qquad \rho_\theta^{\mathbb{C}}(\mathcal{A}, \mathcal{B}) = \max \big\{ \theta(|\alpha|, |\beta|) : \beta(S\mathcal{A})\mathbf{x}^{m-1} = \alpha\mathcal{B}\mathbf{x}^{m-1},$$
$$(\alpha, \beta) \in \mathbb{C}_{1,2}, \mathbf{x} \in \mathbb{C}^n \setminus \{\mathbf{0}\}, |S| = I \big\}.$$

For the same reason, there exist three signature matrices S_1, S_2, and T such that

$$S_1(\mathcal{A}\mathbf{x}^{m-1}) = \big|\mathcal{A}\mathbf{x}^{m-1}\big|, \quad S_2(\mathcal{B}\mathbf{x}^{m-1}) = \big|\mathcal{B}\mathbf{x}^{m-1}\big|, \quad T^*\mathbf{x} = |\mathbf{x}|,$$

so $\left|\beta\mathcal{A}\mathbf{x}^{m-1}\right| = \left|\alpha\mathcal{B}\mathbf{x}^{m-1}\right|$ is equivalent to $|\beta|(S_1\mathcal{A}T^{m-1})|\mathbf{x}|^{m-1} = |\alpha|(S_2\mathcal{B}T^{m-1})|\mathbf{x}|^{m-1}$. Notice that $|\alpha|$, $|\beta|$, and $|\mathbf{x}|$ are all nonnegative. Denote $S = S_2^* S_1$. Thus the sign-complex θ-spectral radius has the third equivalent definition:

(2.12)
$$\rho_\theta^{\mathbb{C}}(\mathcal{A}, \mathcal{B}) = \max\left\{\theta(\alpha, \beta) : \beta(S\mathcal{A}T^{m-1})\mathbf{x}^{m-1} = \alpha(\mathcal{B}T^{m-1})\mathbf{x}^{m-1},\right.$$
$$\left.(\alpha, \beta) \in (\mathbb{R}_+)_{1,2}, \mathbf{x} \in \mathbb{R}_+^n \setminus \{\mathbf{0}\}, |S| = |T| = I\right\}.$$

When $\det(\mathcal{B}) \neq 0$, we can also define the *sign-complex spectral radius* of a regular pair $\{\mathcal{A}, \mathcal{B}\}$ in three equivalent ways, that is,

(2.13)
$$\rho^{\mathbb{C}}(\mathcal{A}, \mathcal{B}) = \max\left\{|\lambda| : \left|\mathcal{A}\mathbf{x}^{m-1}\right| = \left|\lambda\mathcal{B}\mathbf{x}^{m-1}\right|, \lambda \in \mathbb{C}, \mathbf{x} \in \mathbb{C}^n \setminus \{\mathbf{0}\}\right\}$$
$$= \max\left\{|\lambda| : (S\mathcal{A})\mathbf{x}^{m-1} = \lambda\mathcal{B}\mathbf{x}^{m-1}, \lambda \in \mathbb{C}, \mathbf{x} \in \mathbb{C}^n \setminus \{\mathbf{0}\}, |S| = I\right\}$$
$$= \max\left\{\lambda \in \mathbb{R}_+ : (S\mathcal{A}T^{m-1})\mathbf{x}^{m-1} = \lambda(\mathcal{B}T^{m-1})\mathbf{x}^{m-1},\right.$$
$$\left.\mathbf{x} \in \mathbb{R}_+^n \setminus \{\mathbf{0}\}, |S| = |T| = I\right\}.$$

We have the similar relationship that

$$\rho(\mathcal{A}, \mathcal{B}) \leq \rho^{\mathbb{C}}(\mathcal{A}, \mathcal{B}) = \tan\left(\rho_\theta^{\mathbb{C}}(\mathcal{A}, \mathcal{B})\right) \leq \psi(\mathcal{A}, \mathcal{B}).$$

Proposition 2.17. *Let $\{\mathcal{A}, \mathcal{B}\}$ be a regular tensor pair and Q be a nonsingular matrix. Assume that the matrix pair $\{X, Y\}$ is in one of the three situations:*

1. $\{X, Y\} = \{S_1, S_2\}$, *where S_1 and S_2 are two signature matrices;*

2. $\{X, Y\} = \{D, D\}$, *where D is a diagonal matrix;*

3. $\{X, Y\} = \{P, P\}$, *where P is a permutation matrix.*

Then it holds that

$$\rho_\theta^{\mathbb{C}}(\mathcal{A}, \mathcal{B}) = \rho_\theta^{\mathbb{C}}(X\mathcal{A}Q^{m-1}, Y\mathcal{B}Q^{m-1}),$$

and when $\det(\mathcal{B}) \neq 0$, it holds that

$$\rho^{\mathbb{C}}(\mathcal{A}, \mathcal{B}) = \rho^{\mathbb{C}}(X\mathcal{A}Q^{m-1}, Y\mathcal{B}Q^{m-1}).$$

Proof. Notice that if S is a signature matrix, then $Y^{-1}SX$ is still a signature matrix. Then applying (2.11) and (2.13), we obtain the results directly. $\square$

2.5.2 Collatz-Wielandt Formula

The Perron-Frobenius theorem [49, Section 7.3.4] is a well-known result that describes the spectral radius of a nonnegative matrix, and the Collatz-Wielandt formula
$$\rho(A) = \max_{\mathbf{x} \in \mathbb{R}_+^n} \min_{\substack{i=1,2,\ldots,n \\ x_i \neq 0}} \frac{(A\mathbf{x})_i}{x_i} \quad (A \text{ is nonnegative})$$

is one of the most important parts of Perron-Frobenius theorem. Recently, this theorem was extended to the nonnegative tensor case [19, 129, 128]. Rump [108] proposed a generalization of the Perron-Frobenius theorem, mainly of the

Collatz-Wielandt formula, for a general complex matrix employing the sign-complex spectral radius. In this subsection, we further generalize this theorem to the *complex tensor pair* situation.

The following lemma is also used in [108], which indicates that each multivariate polynomial has a root with components having the same absolute value.

Lemma 2.18 ([44]). *For a multivariate polynomial* $p \in \mathbb{C}[z_1, z_2, \ldots, z_n]$, *denote*

$$\mu := \min \left\{ \|\mathbf{z}\|_\infty : p(\mathbf{z}) = 0, \mathbf{z} \in \mathbb{C}^n \right\}.$$

Then there exists some $\mathbf{u} \in \mathbb{C}^n$ *with* $p(\mathbf{u}) = 0$ *and* $|u_i| = \mu$ *for* $i = 1, 2, \ldots, n$.

Applying Lemma 2.18, we can obtain another useful result, which will be used several times.

Lemma 2.19. *Let* $\{\mathcal{A}, \mathcal{B}\}$ *be an* m^{th}-*order* n-*dimensional regular tensor pair. If there exists a vector* $\mathbf{x} \in \mathbb{C}^n \setminus \{\mathbf{0}\}$ *such that* $|\beta_0 \mathcal{A}\mathbf{x}^{m-1}| \geq |\alpha_0 \mathcal{B}\mathbf{x}^{m-1}|$, *then*

$$\rho_\theta^{\mathbb{C}}(\mathcal{A}, \mathcal{B}) \geq \theta(|\alpha_0|, |\beta_0|).$$

Proof. Because $|\beta_0 \mathcal{A}\mathbf{x}^{m-1}| \geq |\alpha_0 \mathcal{B}\mathbf{x}^{m-1}|$, there is a diagonal matrix $D = \text{diag}(d_1, d_2, \ldots, d_n)$ such that $|d_i| \leq 1$ for $i = 1, 2, \ldots, n$ and $\beta_0 D \mathcal{A}\mathbf{x}^{m-1} = \alpha_0 \mathcal{B}\mathbf{x}^{m-1}$. Thus the multivariate polynomial $p(d_1, d_2, \ldots, d_n) := \det(\beta_0 D \mathcal{A} - \alpha_0 \mathcal{B}) = 0$. By Lemma 2.18, there exists another diagonal matrix $\widetilde{D} = \text{diag}(\widetilde{d}_1, \widetilde{d}_2, \ldots, \widetilde{d}_n)$ such that $|\widetilde{d}_i| = \mu \leq 1$ for $i = 1, 2, \ldots, n$ and $p(\widetilde{d}_1, \widetilde{d}_2, \ldots, \widetilde{d}_n) = \det(\beta_0 \widetilde{D} \mathcal{A} - \alpha_0 \mathcal{B}) = 0$.

If $\mu = 0$, then $\rho_\theta^{\mathbb{C}}(\mathcal{A}, \mathcal{B}) = \pi/2$. Thus the result is apparent.

If $\mu \neq 0$, then $\mu^{-1}\widetilde{D}$ is a signature matrix and $\det\left(\beta_0(\mu^{-1}\widetilde{D}\mathcal{A}) - (\mu^{-1}\alpha_0)\mathcal{B}\right) = 0$. Hence

$$\rho_\theta^{\mathbb{C}}(\mathcal{A}, \mathcal{B}) = \rho_\theta^{\mathbb{C}}(\mu^{-1}\widetilde{D}\mathcal{A}, \mathcal{B}) \geq \theta(|\mu^{-1}\alpha_0|, |\beta_0|) \geq \theta(|\alpha_0|, |\beta_0|),$$

by Proposition 2.17 and $|\mu^{-1}| \geq 1$. □

Denote by

$$\theta_0(\mathbf{x}) := \min_{i=1,2,\ldots,n} \theta\left(\left|(\mathcal{A}\mathbf{x}^{m-1})_i\right|, \left|(\mathcal{B}\mathbf{x}^{m-1})_i\right|\right)$$

for any $\mathbf{x} \in \mathbb{C}^n \setminus \{\mathbf{0}\}$. Then we can define two functions on $\mathbb{C}^n \setminus \{\mathbf{0}\}$:

$$\alpha_0(\mathbf{x}) := \sin\left(\theta_0(\mathbf{x})\right), \quad \beta_0(\mathbf{x}) := \cos\left(\theta_0(\mathbf{x})\right).$$

Thus it can be shown that $\beta_0(\mathbf{x})|\mathcal{A}\mathbf{x}^{m-1}| \geq \alpha_0(\mathbf{x})|\mathcal{B}\mathbf{x}^{m-1}|$. By Lemma 2.19, we have

$$\rho_\theta^{\mathbb{C}}(\mathcal{A}, \mathcal{B}) \geq \theta\left(\alpha_0(\mathbf{x}), \beta_0(\mathbf{x})\right) = \min_{i=1,2,\ldots,n} \theta\left(\left|(\mathcal{A}\mathbf{x}^{m-1})_i\right|, \left|(\mathcal{B}\mathbf{x}^{m-1})_i\right|\right)$$

for an arbitrary nonzero complex vector $\mathbf{x}$.

Furthermore, there exists a vector $\mathbf{y}$ such that $\rho_\theta^{\mathbb{C}}(\mathcal{A}, \mathcal{B}) = \theta(|\alpha|, |\beta|)$ and

$$|\beta| \cdot |\mathcal{A}\mathbf{y}^{m-1}| = |\alpha| \cdot |\mathcal{B}\mathbf{y}^{m-1}|.$$

Then $\theta\left(\alpha_0(\mathbf{y}), \beta_0(\mathbf{y})\right) = \theta(|\alpha|, |\beta|) = \rho_\theta^{\mathbb{C}}(\mathcal{A}, \mathcal{B})$. Therefore we obtain the following theorem, which can be regarded as a generalized Collatz-Wielandt formula for complex regular tensor pairs.

Theorem 2.20. *Let $\{\mathcal{A}, \mathcal{B}\}$ be an mth-order n-dimensional regular tensor pair. Then the sign-complex θ-spectral radius has the variational form*

$$(2.14) \qquad \rho_\theta^{\mathbb{C}}(\mathcal{A}, \mathcal{B}) = \max_{\mathbf{x} \in \mathbb{C}^n \setminus \{0\}} \min_{i=1,2,\ldots,n} \theta\Big(\big|(\mathcal{A}\mathbf{x}^{m-1})_i\big|, \big|(\mathcal{B}\mathbf{x}^{m-1})_i\big|\Big).$$

When $\det(\mathcal{B}) \neq 0$, the above generalized Collatz-Wielandt formula can be rewritten into the following more familiar form, which leads to the single tensor case more directly.

Theorem 2.21. *Let $\{\mathcal{A}, \mathcal{B}\}$ be an mth-order n-dimensional regular tensor pair. Assume that $\det(\mathcal{B}) \neq 0$. Then the sign-complex spectral radius has the variational form*

$$(2.15) \qquad \rho^{\mathbb{C}}(\mathcal{A}, \mathcal{B}) = \max_{\substack{\mathbf{x} \in \mathbb{C}^n \setminus \{0\}}} \min_{\substack{i=1,2,\ldots,n \\ (\mathcal{B}\mathbf{x}^{m-1})_i \neq 0}} \left| \frac{(\mathcal{A}\mathbf{x}^{m-1})_i}{(\mathcal{B}\mathbf{x}^{m-1})_i} \right|.$$

The last result in this subsection studies the performance of the sign-complex θ-spectral radius under the componentwise changes. Use $\mathcal{E}_{i_1, i_2, \ldots, i_m}$ to denote a tensor with a 1 in $(i_1, i_2, \ldots, i_m)$ position and 0 elsewhere.

Proposition 2.22. *Let $\{\mathcal{A}, \mathcal{B}\}$ be an mth-order n-dimensional regular tensor pair. Then for all $i_1, i_2, \ldots, i_m \in \{1, 2, \ldots, n\}$, there exist half spaces $\mathscr{H}_1$ and $\mathscr{H}_2$ in $\mathbb{C}$ such that*

$$\rho_\theta^{\mathbb{C}}(\mathcal{A} + h_1 \mathcal{E}_{i_1, i_2, \ldots, i_m}, \mathcal{B}) \geq \rho_\theta^{\mathbb{C}}(\mathcal{A}, \mathcal{B}) \text{ for all } h_1 \in \mathscr{H}_1,$$

and

$$\rho_\theta^{\mathbb{C}}(\mathcal{A}, \mathcal{B} - h_2 \mathcal{E}_{i_1, i_2, \ldots, i_m}) \geq \rho_\theta^{\mathbb{C}}(\mathcal{A}, \mathcal{B}) \text{ for all } h_2 \in \mathscr{H}_2.$$

Proof. Let the vector $\mathbf{x}$ satisfy $\big|\beta_0 \mathcal{A}\mathbf{x}^{m-1}\big| = \big|\alpha_0 \mathcal{B}\mathbf{x}^{m-1}\big|$, where $\theta(|\alpha_0|, |\beta_0|) = \rho_\theta^{\mathbb{C}}(\mathcal{A}, \mathcal{B})$. Then there is a half space $\mathscr{H}_1$ in $\mathbb{C}$ with all $h_1 \in \mathscr{H}_1$ such that

$$h_1 x_{i_2} \ldots x_{i_m} = t e^{\imath\varphi}, \ t \geq 0, \ \varphi \in \left[\phi - \frac{\pi}{2}, \phi + \frac{\pi}{2} \right],$$

where $\imath = \sqrt{-1}$ and ϕ is the argument of $(\mathcal{A}\mathbf{x}^{m-1})_{i_1}$. Thus it can be guaranteed that

$$\big|(\mathcal{A}\mathbf{x}^{m-1})_{i_1} + h_1 x_{i_2} \ldots x_{i_m}\big| \geq \big|(\mathcal{A}\mathbf{x}^{m-1})_{i_1}\big|,$$

which indicates that

$$\big|\beta_0(\mathcal{A} + h_1 \mathcal{E}_{i_1, i_2, \ldots, i_m})\mathbf{x}^{m-1}\big| \geq \big|\beta_0 \mathcal{A}\mathbf{x}^{m-1}\big| = \big|\alpha_0 \mathcal{B}\mathbf{x}^{m-1}\big|.$$

Then by Lemma 2.19, we obtain the first statement.

The second statement can be proved in the same way. $\qquad\square$

2.5.3 Properties for Single Tensors

Particularly, we can derive the sign-complex spectral radius for a single complex tensor by taking $\mathcal{B}$ as the unit tensor $\mathcal{I}$, that is,

$$(2.16)$$
$$\rho^{\mathbb{C}}(\mathcal{A}) = \max \big\{ |\lambda| : \big|\mathcal{A}\mathbf{x}^{m-1}\big| = \big|\lambda \mathbf{x}^{[m-1]}\big|, \lambda \in \mathbb{C}, \mathbf{x} \in \mathbb{C}^n \setminus \{0\} \big\}$$
$$= \max \big\{ |\lambda| : (S\mathcal{A})\mathbf{x}^{m-1} = \lambda \mathbf{x}^{[m-1]}, \lambda \in \mathbb{C}, \mathbf{x} \in \mathbb{C}^n \setminus \{0\}, |S| = \mathcal{I} \big\}$$
$$= \max \big\{ \lambda \in \mathbb{R}_+ : (S\mathcal{A}T^{m-1})\mathbf{x}^{m-1} = \lambda \mathbf{x}^{[m-1]}, \mathbf{x} \in \mathbb{R}_+^n \setminus \{0\}, |S| = |T| = \mathcal{I} \big\}.$$

Note that the third equivalent definition is modified, since we have $\mathcal{I}T^{m-1} = (T^{m-1})\mathcal{I}$ for a diagonal matrix T.

As a corollary of Theorem 2.21, the following result can be treated as a generalized Collatz-Wielandt formula for a single complex tensor.

Corollary 2.23. *Let $\mathcal{A}$ be an mth-order n-dimensional complex tensor. Then the sign-complex spectral radius has the variational form*

$$(2.17) \qquad \rho^{\mathbb{C}}(\mathcal{A}) = \max_{\mathbf{x} \in \mathbb{C}^n \setminus \{0\}} \min_{\substack{i=1,2,\ldots,n \\ x_i \neq 0}} \left| \frac{(\mathcal{A}\mathbf{x}^{m-1})_i}{x_i^{m-1}} \right|.$$

When $\mathcal{A}$ is nonnegative, we show that the sign-complex spectral radius exactly equals to the spectral radius. On the one hand, the inequality $\rho(\mathcal{A}) \leq \rho^{\mathbb{C}}(\mathcal{A})$ is obvious. On the other hand, we denote $\gamma = \rho^{\mathbb{C}}(\mathcal{A})$, then there exists a vector $\mathbf{x} \in \mathbb{C}^n \setminus \{0\}$ such that $\mathcal{A}|\mathbf{x}|^{m-1} \geq |\mathcal{A}\mathbf{x}^{m-1}| = |\gamma\mathbf{x}^{[m-1]}| = \gamma|\mathbf{x}|^{[m-1]}$. Then $\rho(\mathcal{A}) \geq \gamma = \rho^{\mathbb{C}}(\mathcal{A})$ by [129, Theorem 5.3]. We prove that $\rho(\mathcal{A}) = \rho^{\mathbb{C}}(\mathcal{A})$ for a nonnegative tensor $\mathcal{A}$. Hence Corollary 2.23 degenerates into

$$\rho(\mathcal{A}) = \max_{\mathbf{x} \in \mathbb{R}^n_+ \setminus \{0\}} \min_{\substack{i=1,2,\ldots,n \\ x_i > 0}} \frac{(\mathcal{A}\mathbf{x}^{m-1})_i}{x_i^{m-1}},$$

when $\mathcal{A}$ is nonnegative, which is exactly the result of [129, Theorem 5.3]. By the same reason, the following corollary of Proposition 2.22 can be regarded as a generalization of [129, Lemma 3.5].

Corollary 2.24. *Let $\mathcal{A}$ be an mth-order n-dimensional complex tensor. Then for all $i_1, i_2, \ldots, i_m \in \{1, 2, \ldots, n\}$, there exists a half space $\mathscr{H}$ in $\mathbb{C}$ such that*

$$\rho^{\mathbb{C}}(\mathcal{A} + h\mathcal{E}_{i_1,i_2,\ldots,i_m}) \geq \rho^{\mathbb{C}}(\mathcal{A}) \text{ for all } h \in \mathscr{H}.$$

Apart from these results derived from the tensor pair case, we also have some special ones about the sign-complex spectral radius of a single tensor.

Proposition 2.25. *Let $\mathcal{A}$ be an mth-order n-dimensional complex tensor. Then*

1. *$\rho(\mathcal{A}) \leq \rho^{\mathbb{C}}(\mathcal{A}) \leq \rho^{\mathbb{C}}(|\mathcal{A}|) = \rho(|\mathcal{A}|)$;*

2. *$\rho^{\mathbb{C}}(\mathcal{A}) \leq \|\mathcal{A}\|_\infty$;*

3. *$\rho^{\mathbb{C}}(\mathcal{A}) \geq \rho^{\mathbb{C}}(\mathcal{A}_J)$, where $\mathcal{A}_J$ is the principal subtensor of the tensor $\mathcal{A}$ with the index set J.*

Proof. We denote $\gamma = \rho^{\mathbb{C}}(\mathcal{A})$ in this proof.

1. We only need to prove that $\rho^{\mathbb{C}}(\mathcal{A}) \leq \rho^{\mathbb{C}}(|\mathcal{A}|)$. Let $\mathbf{x}$ satisfy that $|\mathcal{A}\mathbf{x}^{m-1}| = |\gamma\mathbf{x}^{[m-1]}|$. Then $|\mathcal{A}||\mathbf{x}|^{m-1} \geq |\mathcal{A}\mathbf{x}^{m-1}| = \gamma|\mathbf{x}|^{[m-1]}$. Therefore $\rho^{\mathbb{C}}(|\mathcal{A}|) \geq \gamma$ by Corollary 2.23.

2. From the second definition of $\rho^{\mathbb{C}}(\mathcal{A})$, there is a signature matrix S and a nonzero vector $\mathbf{x}$ such that $S\mathcal{A}\mathbf{x}^{m-1} = \gamma\mathbf{x}^{[m-1]}$. Recall that $\|\mathbf{x}^{[m-1]}\|_\infty = \|\mathbf{x}^{\otimes(m-1)}\|_\infty = \|\mathbf{x}\|_\infty^{m-1}$. Then

$$\gamma\|\mathbf{x}\|_\infty^{m-1} \leq \|(S\mathcal{A})_{(1)}\|_\infty\|\mathbf{x}^{\otimes(m-1)}\|_\infty = \|S \cdot \mathcal{A}_{(1)}\|_\infty\|\mathbf{x}\|_\infty^{m-1}$$
$$\leq \|S\|_\infty\|\mathcal{A}_{(1)}\|_\infty\|\mathbf{x}\|_\infty^{m-1}.$$

Since S is a signature matrix, then $\|S\|_\infty = 1$. Therefore we obtain that $\gamma \leq \|\mathcal{A}\|_\infty$.

3. Let $\mathbf{z} \in \mathbb{C}^k$ satisfy $\left| \mathcal{A}_J \mathbf{z}^{m-1} \right| = \left| \gamma_J \mathbf{z}^{[m-1]} \right|$, where $\gamma_J = \rho^{\mathbb{C}}(\mathcal{A}_J)$ and k is the size of J. Use $\mathrm{aug}(\mathbf{z})$ to denote the augmentation of $\mathbf{z}$ by zeros into $\mathbb{C}^n$. Then

$$\left| \mathcal{A} \mathrm{aug}(\mathbf{z})^{m-1} \right| \geq \left| \mathrm{aug}(\mathcal{A}_J \mathbf{z}^{m-1}) \right| = \left| \mathrm{aug}(\gamma_J \mathbf{z}^{[m-1]}) \right| = \left| \gamma_J \mathrm{aug}(\mathbf{z})^{[m-1]} \right|.$$

Therefore $\rho^{\mathbb{C}}(\mathcal{A}) \geq \gamma_J$ by Corollary 2.23.

$\square$

From this proposition, we can see the significance of the Collatz-Wielandt formula. Since all the eigenvalues of a tensor are generally very hard to compute and the sign-complex spectral radius is smaller than other easy estimates of the spectral radius, such as $\rho(|\mathcal{A}|)$ and $\|\mathcal{A}\|_\infty$, the sign-complex spectral radius is a tight and computable estimation.

2.5.4 The Componentwise Distance to Singularity

As for a nonsingular tensor, that is, its determinant is not zero, we wonder how far it is from singularity. It is well-known that the normwise distance of a nonsingular matrix A to singularity is its condition number $\kappa(A) = \|A\|_2 \|A^{-1}\|_2$ [49, Section 2.6]; Rump [108] stated that the componentwise distance with the weight matrix E is $\left\{ \max_{|S|=I} \rho^{\mathbb{C}}(A^{-1} S E) \right\}^{-1}$. We will discuss the componentwise distance of a nonsingular tensor to singularity in this subsection.

Let $\mathcal{A}$ be a nonsingular tensor and $\mathcal{E}$ be a nonnegative weight tensor. The componentwise distance of $\mathcal{A}$ to singularity corresponding to $\mathcal{E}$ is defined as

$$d_{\mathcal{E}}^{\mathbb{C}}(\mathcal{A}) := \min \left\{ \delta \in \mathbb{R}_+ : \det \left(\mathcal{A} + \widetilde{\mathcal{E}} \right) = 0, \left| \widetilde{\mathcal{E}} \right| \leq \delta \mathcal{E} \right\},$$

which is equivalent to

$$d_{\mathcal{E}}^{\mathbb{C}}(\mathcal{A}) = \min \left\{ \delta \in \mathbb{R}_+ : \left(\mathcal{A} + \widetilde{\mathcal{E}} \right) \mathbf{x}^{m-1} = \mathbf{0}, \mathbf{x} \in \mathbb{C}^n \setminus \{\mathbf{0}\}, \left| \widetilde{\mathcal{E}} \right| \leq \delta \mathcal{E} \right\}.$$

We use the sign-complex spectral radius to measure this distance. However, we need a lemma first.

Lemma 2.26. *Let $\mathcal{A}$ be a complex tensor, $\mathcal{E}$ be a nonnegative tensor, $\mathbf{b}$ be a complex vector, and $\mathbf{d}$ be a nonnegative vector. Define*

$$\Sigma := \left\{ \mathbf{x} \in \mathbb{C}^n : \left(\mathcal{A} + \widetilde{\mathcal{E}} \right) \mathbf{x}^{m-1} = \mathbf{b} + \widetilde{\mathbf{d}}, \left| \widetilde{\mathcal{E}} \right| \leq \mathcal{E}, \left| \widetilde{\mathbf{d}} \right| \leq \mathbf{d} \right\}.$$

Then $\Sigma = \left\{ \mathbf{x} \in \mathbb{C}^n : \left| \mathcal{A} \mathbf{x}^{m-1} - \mathbf{b} \right| \leq \mathcal{E}|\mathbf{x}|^{m-1} + \mathbf{d} \right\}$.

Proof. On the one hand, if $\mathbf{x} \in \Sigma$, then $\left| \mathcal{A} \mathbf{x}^{m-1} - \mathbf{b} \right| = \left| -\widetilde{\mathcal{E}} \mathbf{x}^{m-1} + \widetilde{\mathbf{d}} \right| \leq \mathcal{E}|\mathbf{x}|^{m-1} + \mathbf{d}$.

On the other hand, if $\left| \mathcal{A} \mathbf{x}^{m-1} - \mathbf{b} \right| \leq \mathcal{E}|\mathbf{x}|^{m-1} + \mathbf{d}$, then there exist two signature matrices S_1 and S_2 such that $S_1 \left(\mathcal{A} \mathbf{x}^{m-1} - \mathbf{b} \right) = \left| \mathcal{A} \mathbf{x}^{m-1} - \mathbf{b} \right|$ and $S_2 \mathbf{x} = |\mathbf{x}|$, which indicates that $S_1 \left(\mathcal{A} \mathbf{x}^{m-1} - \mathbf{b} \right) \leq \mathcal{E} S_2^{m-1} \mathbf{x}^{m-1} + \mathbf{d}$. Thus there is a diagonal matrix D with $|D| \leq I$ such that $\mathcal{A} \mathbf{x}^{m-1} - \mathbf{b} = S_1^* D \mathcal{E} S_2^{m-1} \mathbf{x}^{m-1} + S_1^* D \mathbf{d}$. Take $\widetilde{\mathcal{E}} = -S_1^* D \mathcal{E} S_2^{m-1}$ and $\widetilde{\mathbf{d}} = S_1^* D \mathbf{d}$. Then it is obvious that $\mathbf{x} \in \Sigma$. $\square$

Hence from this lemma, we have the third way to define $d_{\mathcal{E}}^{\mathbb{C}}(\mathcal{A})$, that is,

$$d_{\mathcal{E}}^{\mathbb{C}}(\mathcal{A}) = \min\left\{\delta \in \mathbb{R}_+ : \left|\mathcal{A}\mathbf{x}^{m-1}\right| \le \delta\mathcal{E}|\mathbf{x}|^{m-1}, \mathbf{x} \in \mathbb{C}^n \setminus \{\mathbf{0}\}\right\}.$$

We have the following theorem on the distance to singularity.

Theorem 2.27. *Let $\mathcal{A}$ be a nonsingular tensor and $\mathcal{E}$ be a nonnegative weight tensor. Then the componentwise distance of $\mathcal{A}$ to the nearest singular tensor corresponding to $\mathcal{E}$ is*

$$(2.18) \qquad d_{\mathcal{E}}^{\mathbb{C}}(\mathcal{A}) = \left\{\max_{|S|=I} \rho^{\mathbb{C}}(\mathcal{E}S^{m-1}, \mathcal{A})\right\}^{-1}.$$

Proof. Denote $\delta_0 = d_{\mathcal{E}}^{\mathbb{C}}(\mathcal{A})$. There is a nonzero vector $\mathbf{x}$ and two signature matrices S_1 and S_2 satisfying $S_1\mathcal{A}S_2^{m-1}|\mathbf{x}|^{m-1} = \left|\mathcal{A}\mathbf{x}^{m-1}\right| \le \delta_0\mathcal{E}|\mathbf{x}|^{m-1}$. This implies that there is a diagonal matrix D with $|D| \le I$ such that $S_1\mathcal{A}S_2^{m-1}|\mathbf{x}|^{m-1} = \delta_0 D\mathcal{E}|\mathbf{x}|^{m-1}$, which further implies that $\det\left(S_1\mathcal{A}S_2^{m-1} - \delta_0 D\mathcal{E}\right) = 0$. Thus Lemma 2.18 ensures the existence of another diagonal matrix $\widetilde{D}$ with $|\widetilde{D}| = \mu I \le I$ such that $\det\left(S_1\mathcal{A}S_2^{m-1} - \delta_0\widetilde{D}\mathcal{E}\right) = 0$. Let $\mathbf{y}$ be a nonzero vector such that $S_1\mathcal{A}S_2^{m-1}\mathbf{y}^{m-1} = \delta_0\widetilde{D}\mathcal{E}\mathbf{y}^{m-1}$, which indicates that $\left|\mathcal{A}\mathbf{z}^{m-1}\right| = \left|\mathcal{A}S_2^{m-1}\mathbf{y}^{m-1}\right| = (\delta_0\mu)\left|\mathcal{E}\mathbf{y}^{m-1}\right| \le (\delta_0\mu)\mathcal{E}|\mathbf{y}|^{m-1} = (\delta_0\mu)\mathcal{E}|\mathbf{z}|^{m-1}$, where $\mathbf{z} = S_2\mathbf{y}$. By the definition of δ_0, we know that $\mu = 1$, then $\left|\mathcal{E}(S_2^*)^{m-1}\mathbf{z}^{m-1}\right| = \left|\delta_0^{-1}\mathcal{A}\mathbf{z}^{m-1}\right|$. Therefore we have $\rho^{\mathbb{C}}\left(\mathcal{E}(S_2^*)^{m-1}, \mathcal{A}\right) \ge \delta_0^{-1}$, which implies that

$$d_{\mathcal{E}}^{\mathbb{C}}(\mathcal{A}) = \delta_0 \ge \left\{\max_{|S|=I} \rho^{\mathbb{C}}(\mathcal{E}S^{m-1}, \mathcal{A})\right\}^{-1}.$$

The reverse inequality is easy to verify, and we obtain the final equality. $\qquad\square$

Example 2.4. We present a simple example of $\mathcal{M}$-tensors [38, 133] with diagonal perturbations. Given an $\mathcal{M}$-tensor $\mathcal{A} = s\mathcal{I} - \mathcal{B}$, where $\mathcal{B}$ is a nonnegative tensor and $s > \rho(\mathcal{B})$. We wonder how large a complex perturbation on the diagonal can be while preserving the nonsingularity. Thus we set the weight tensor $\mathcal{E} = \mathcal{I}$. According to Theorems 2.21 and 2.27, we know that the distance of $\mathcal{A}$ to singularity is

$$d_{\mathcal{E}}^{\mathbb{C}}(\mathcal{A}) = \left\{\max_{|S|=I} \max_{\substack{\mathbf{x}\in\mathbb{C}^n\setminus\{\mathbf{0}\}}} \min_{\substack{i=1,2,\ldots,n \\ (\mathcal{A}\mathbf{x}^{m-1})_i \neq 0}} \left|\frac{(\mathcal{I}S^{m-1}\mathbf{x}^{m-1})_i}{(\mathcal{A}\mathbf{x}^{m-1})_i}\right|\right\}^{-1}$$

$$= \left\{\max_{\substack{\mathbf{x}\in\mathbb{C}^n\setminus\{\mathbf{0}\}}} \min_{\substack{i=1,2,\ldots,n \\ (\mathcal{A}\mathbf{x}^{m-1})_i \neq 0}} \frac{|x_i^{m-1}|}{|sx_i^{m-1} - (\mathcal{B}\mathbf{x}^{m-1})_i|}\right\}^{-1}$$

$$= \min_{\substack{\mathbf{x}\in\mathbb{C}^n\setminus\{\mathbf{0}\}}} \max_{\substack{i=1,2,\ldots,n \\ x_i \neq 0}} \left|s - \frac{(\mathcal{B}\mathbf{x}^{m-1})_i}{x_i^{m-1}}\right|$$

$$\ge s - \max_{\substack{\mathbf{x}\in\mathbb{C}^n\setminus\{\mathbf{0}\}}} \min_{\substack{i=1,2,\ldots,n \\ x_i \neq 0}} \left|\frac{(\mathcal{B}\mathbf{x}^{m-1})_i}{x_i^{m-1}}\right|$$

$$= s - \rho^{\mathbb{C}}(\mathcal{B}) = s - \rho(\mathcal{B}).$$

The equality can be attained by taking $\mathbf{x}$ as an eigenvector of $\mathcal{B}$ corresponding to the eigenvalue $\rho(\mathcal{B})$. Therefore $d_{\mathcal{E}}^{\mathbb{C}}(\mathcal{A}) = s - \rho(\mathcal{B})$, which is realized with $\rho(\mathcal{B})\mathcal{I} - \mathcal{B}$ being a singular $\mathcal{M}$-tensor.

2.5.5 Bauer-Fike Theorem

The Bauer-Fike theorem [7] is a well-known result for the perturbation of the eigenvalues of matrices. The classical Bauer-Fike theorem reveals that the nearest distance of the perturbed eigenvalues to the original spectrum is bounded by $\|X\|\,\|X^{-1}\|\,\|E\|$, where X consists of the eigenvectors and E is the perturbation matrix. Shi and Wei [109] sharpened the bound into $\rho^{\mathbb{C}}(X^{-1}EX)$. We now extend these results to diagonalizable tensor pairs defined in Section 2.3.3.

Let $\{\mathcal{A},\mathcal{B}\}$ be a diagonalizable regular tensor pair, and P, Q be two nonsingular matrices such that

$$\mathcal{C} = P^{-1}\mathcal{A}Q^{m-1} \quad \text{and} \quad \mathcal{D} = P^{-1}\mathcal{B}Q^{m-1}$$

are diagonal tensors with the diagonals $(c_1, c_2, \ldots, c_n)$ and $(d_1, d_2, \ldots, d_n)$, respectively. Assume that $\{\widetilde{\mathcal{A}}, \widetilde{\mathcal{B}}\} = \{\mathcal{A}+\mathcal{E}, \mathcal{B}+\mathcal{F}\}$ is another regular tensor pair, and (α, β) is an eigenvalue of $\{\widetilde{\mathcal{A}}, \widetilde{\mathcal{B}}\}$. Thus there is a vector $\mathbf{x} \in \mathbb{C}^n \setminus \{\mathbf{0}\}$ such that

$$\beta(\mathcal{A} + \mathcal{E})\mathbf{x}^{m-1} = \alpha(\mathcal{B} + \mathcal{F})\mathbf{x}^{m-1},$$

which indicates that

$$(\beta\mathcal{C} - \alpha\mathcal{D})\mathbf{y}^{m-1} = P^{-1}(\alpha\mathcal{F} - \beta\mathcal{E})Q^{m-1}\mathbf{y}^{m-1},$$

where $\mathbf{y} = Q^{-1}\mathbf{x}$. Without loss of generality, we can assume that (α, β) is not an eigenvalue of $\{\mathcal{A},\mathcal{B}\}$. Then the diagonal entries of $\beta\mathcal{C} - \alpha\mathcal{D}$ are nonzero, thus

$$\left|P^{-1}(\beta\mathcal{E}-\alpha\mathcal{F})Q^{m-1}\mathbf{y}^{m-1}\right| = \left|(\beta\mathcal{C}-\alpha\mathcal{D})\mathbf{y}^{m-1}\right| \geq \left(\min_{i=1,2,\ldots,n}|\beta c_i-\alpha d_i|\right)\cdot|\mathbf{y}|^{[m-1]}.$$

By Corollary 2.23, we have

$$\min_{i=1,2,\ldots,n}|\beta c_i - \alpha d_i| \leq \rho^{\mathbb{C}}\left(P^{-1}(\beta\mathcal{E} - \alpha\mathcal{F})Q^{m-1}\right).$$

Nonetheless, this is not a satisfactory result, since there are still α and β on the right-hand side. We have two alternatives to avoid this disadvantage.

The first way is to strengthen the condition. We further assume that $\mathcal{B} = \widetilde{\mathcal{B}}$ is nonsingular. This is exactly the situation when we consider the single tensor case by taking $\mathcal{B} = \mathcal{I}$. Thus we have the following sharper bound with a stronger condition, which can be regarded as a generalization of [109, Theorem 2.8].

Theorem 2.28. *Let $\{\mathcal{A},\mathcal{B}\}$ be a diagonalizable regular tensor pair with $\det(\mathcal{B}) \neq 0$, and P, Q be two nonsingular matrices such that $\mathcal{C} = P^{-1}\mathcal{A}Q^{m-1}$ and $\mathcal{D} = P^{-1}\mathcal{B}Q^{m-1}$ are diagonal tensors with diagonals $(c_1, c_2, \ldots, c_n)$ and $(d_1, d_2, \ldots, d_n)$, respectively. Assume that $\{\mathcal{A}+\mathcal{E},\mathcal{B}\}$ is another regular tensor pair and λ is one of its eigenvalues. Then*

$$(2.19) \qquad \min_{i=1,2,\ldots,n}|\lambda - c_i/d_i| \leq \rho^{\mathbb{C}}(P^{-1}\mathcal{E}Q^{m-1}).$$

The second way is to relax the result. By Proposition 2.25, we have

$$
\begin{aligned}
\rho^{\mathbb{C}}\left(P^{-1}(\beta\mathcal{E} - \alpha\mathcal{F})Q^{m-1}\right) &\leq \left\|P^{-1}(\beta\mathcal{E} - \alpha\mathcal{F})Q^{m-1}\right\|_{\infty} \\
&= \left\|\left(P^{-1}(\beta\mathcal{E} - \alpha\mathcal{F})Q^{m-1}\right)_{(1)}\right\|_{\infty} \\
&= \left\|P^{-1}(\beta\mathcal{E} - \alpha\mathcal{F})_{(1)}Q^{\otimes(m-1)}\right\|_{\infty} \\
&\leq \|P^{-1}\|_{\infty}\|\beta\mathcal{E} - \alpha\mathcal{F}\|_{\infty}\|Q\|_{\infty}^{m-1}.
\end{aligned}
$$

Applying the same technique in Section 2.3.4, we can derive

$$\|\beta\mathcal{E} - \alpha\mathcal{F}\|_\infty = \max_{i=1,2,\ldots,n} \sum_{i_2,i_3,\ldots,i_m=1}^{n} |\beta e_{ii_2\ldots i_m} - \alpha f_{ii_2\ldots i_m}|$$

$$\leq \sqrt{|\alpha|^2 + |\beta|^2}.$$

$$\max_{i=1,2,\ldots,n} \sqrt{\left(\sum_{i_2,i_3,\ldots,i_m=1}^{n} |e_{ii_2\ldots i_m}|\right)^2 + \left(\sum_{i_2,i_3,\ldots,i_m=1}^{n} |f_{ii_2\ldots i_m}|\right)^2}.$$

Notice that it can always be satisfied that $|c_i|^2 + |d_i|^2 = 1$ for all $i = 1, 2, \ldots, n$ by adjusting the matrix P. Hence we obtain the following generalization of the Bauer-Fike theorem [113, Section 6.2.3].

Theorem 2.29. *Let $\{\mathcal{A}, \mathcal{B}\}$ be a diagonalizable regular tensor pair, and P, Q be two nonsingular matrices such that $\mathcal{C} = P^{-1}\mathcal{A}Q^{m-1}$ and $\mathcal{D} = P^{-1}\mathcal{B}Q^{m-1}$ are diagonal tensors with the diagonals $(c_1, c_2, \ldots, c_n)$ and $(d_1, d_2, \ldots, d_n)$, respectively, with $|c_i|^2 + |d_i|^2 = 1$ for all i. Assume that $\{\mathcal{A} + \mathcal{E}, \mathcal{B} + \mathcal{F}\}$ is another regular tensor pair, and (α, β) is one of its eigenvalues. Then*

$$(2.20) \qquad \min_{i=1,2,\ldots,n} \mathrm{chord}\big((\alpha, \beta), (c_i, d_i)\big) \leq \|P^{-1}\|_\infty \|Q\|_\infty^{m-1} \|\{\mathcal{E}, \mathcal{F}\}\|_\infty,$$

where
(2.21)

$$\|\{\mathcal{E}, \mathcal{F}\}\|_\infty := \max_{i=1,2,\ldots,n} \sqrt{\left(\sum_{i_2,i_3,\ldots,i_m=1}^{n} |e_{ii_2\ldots i_m}|\right)^2 + \left(\sum_{i_2,i_3,\ldots,i_m=1}^{n} |f_{ii_2\ldots i_m}|\right)^2}.$$

2.6   An Illustrative Example

We investigate the spectral theory of the (R-)regular tensor pairs in this chapter. It is proved that the number of the eigenvalues of an mth-order n-dimensional tensor pair is $n(m-1)^{n-1}$. The perturbations of the generalized spectrum are also discussed. Moreover, we summarize the generalizations proposed in this chapter of some famous results briefly, without elaborating on notations. Let $\{\mathcal{A}, \mathcal{B}\}$ be an mth-order n-dimensional tensor pair.

Gershgorin theorem. If $\{\mathcal{A}, \mathcal{B}\}$ is regular, and $(a_{ii\ldots i}, b_{ii\ldots i}) \neq (0,0)$ for all $i = 1, 2, \ldots, n$, then

$$\lambda(\mathcal{A}, \mathcal{B}) \subseteq \bigcup_{i=1}^{n} \Big\{ (\alpha, \beta) \in \mathbb{C}_{1,2} : \mathrm{chord}\big((\alpha, \beta), (a_{ii\ldots i}, b_{ii\ldots i})\big) \leq \gamma_i \Big\}.$$

Rayleigh-Ritz theorem. If $\{\mathcal{A}, \mathcal{B}\}$ is a real symmetric-definite tensor pair, then

$$\lambda_{\max} = \max_{\mathbf{x} \in \mathbb{R}^n \setminus \{0\}} \frac{\mathcal{A}\mathbf{x}^m}{\mathcal{B}\mathbf{x}^m}, \quad \lambda_{\min} = \min_{\mathbf{x} \in \mathbb{R}^n \setminus \{0\}} \frac{\mathcal{A}\mathbf{x}^m}{\mathcal{B}\mathbf{x}^m}.$$

Collatz-Wielandt formula. If $\{\mathcal{A}, \mathcal{B}\}$ is regular, then

$$\rho_\theta^{\mathbb{C}}(\mathcal{A}, \mathcal{B}) = \max_{\mathbf{x} \in \mathbb{C}^n \setminus \{0\}} \min_{i=1,2,\ldots,n} \theta\Big(\big|(\mathcal{A}\mathbf{x}^{m-1})_i\big|, \big|(\mathcal{B}\mathbf{x}^{m-1})_i\big|\Big).$$

Bauer-Fike theorem. If $\{\mathcal{A}, \mathcal{B}\}$ and $\{\mathcal{A}+\mathcal{E}, \mathcal{B}+\mathcal{F}\}$ are regular, $\mathcal{C} = P^{-1}\mathcal{A}Q^{m-1}$ and $\mathcal{D} = P^{-1}\mathcal{B}Q^{m-1}$ are diagonal with $|c_{ii...i}|^2 + |d_{ii...i}|^2 = 1$ for all i, and (α, β) is an eigenvalue of $\{\mathcal{A} + \mathcal{E}, \mathcal{B} + \mathcal{F}\}$, then

$$\min_{i=1,2,...,n} \text{chord}\big((\alpha, \beta), (c_{ii...i}, d_{ii...i})\big) \leq \|P^{-1}\|_\infty \|Q\|_\infty^{m-1} \|\{\mathcal{E}, \mathcal{F}\}\|_\infty.$$

We shall employ a numerical example to illustrate some of the results in this chapter. The 4th-order 3-dimensional symmetric tensor $\mathcal{A}$ comes from Example 5.1 in Kolda and Mayo [68] and is defined by the unique elements

$$
\begin{aligned}
a_{1111} &= 0.2883, & a_{1112} &= -0.0031, & a_{1113} &= 0.1973, & a_{1122} &= -0.2485, \\
a_{1123} &= -0.2939, & a_{1133} &= 0.3847, & a_{1222} &= 0.2972, & a_{1223} &= 0.1862, \\
a_{1233} &= 0.0919, & a_{1333} &= -0.3619, & a_{2222} &= 0.1241, & a_{2223} &= -0.3420, \\
a_{2233} &= 0.2127, & a_{2333} &= 0.2727, & a_{3333} &= -0.3054.
\end{aligned}
$$

The positive definite tensor $\mathcal{B}$ is taken as the tensor $\mathcal{Z}$ in Example 2.2, which satisfies $\mathcal{B}\mathbf{x}^4 = (\mathbf{x}^T\mathbf{x})^2$ and $\mathcal{B}\mathbf{x}^3 = \mathbf{x} \cdot (\mathbf{x}^T\mathbf{x})$. Consider the generalized tensor eigenvalue problem

$$\beta\mathcal{A}\mathbf{x}^3 = \alpha\mathcal{B}\mathbf{x}^3.$$

As stated before, the determinant of $\mathcal{B}$ is zero because there exists some nonzero $\mathbf{v} \in \mathbb{C}^3$ such that $\mathbf{v}^T\mathbf{v} = 0$. One can check, at least numerically, that there is no such nonzero vector $\mathbf{v}$ such that $\mathcal{A}\mathbf{v}^3 = \mathbf{0}$. Thus the tensor pair $\{\mathcal{A}, \mathcal{B}\}$ is a regular pair. According to Theorem 2.1, this tensor pair should have 27 eigenvalues counting multiplicity. When by the function NSolve in Mathematica we solve the problem

$$\begin{cases} \mathcal{A}\mathbf{x}^3 = \lambda\mathcal{B}\mathbf{x}^3, \\ \mathbf{x}^T\mathbf{x} = 1, \end{cases}$$

we obtain 13 distinct λ, 11 of which are real

$$
\begin{array}{llllll}
0.262803, & -0.562917, & 0.363312, & 0.889327, & 0.510489, & 0.816891, \\
0.173475, & 0.243342, & -1.09531, & 0.268256, & -0.0450464, &
\end{array}
$$

and 2 are complex

$$0.676441 + 0.00144494\imath, \quad 0.676441 - 0.00144494\imath.$$

These eigenvalues correspond to the eigenvalues $(\lambda, 1)$ in homogenous form. This indicates that the multiplicity of the eigenvalue $(1, 0)$ with eigenvectors satisfying $\mathbf{x}^T\mathbf{x} = 0$ is 14. To confirm this, we transfor as in the proof of Theorem 2.1 with $\gamma = \delta = 1/\sqrt{2}$. If we solve $\widetilde{\mathcal{A}}\mathbf{x}^{m-1} = \widetilde{\lambda}\widetilde{\mathcal{B}}\mathbf{x}^{m-1}$, we obtain 27 eigenvalues, 13 of which correspond to the above λ and 14 of which are 1 corresponding to $(1, 0)$ eigenvalues of $\{\mathcal{A}, \mathcal{B}\}$.

Following Theorem 2.4, we can compute the three Gershgorin circles of $\{\mathcal{A}, \mathcal{B}\}$,

Center	Radius
$(0.2883, 1)$	5.8564
$(0.1244, 1)$	6.1369
$(-0.3054, 1)$	6.2551

Since neither $\mathcal{A}$ nor $\mathcal{B}$ is diagonally dominant so all the circles have radii greater than 1 and cover the whole space $\mathbb{C}_{1,2}$, thus the Gershgorim circle theorem is trivial for this tensor pair.

For evaluating the accuracy of these computed eigenpairs, we compute the backward errors. By (2.4)–(2.5), we have the relative normwise backward errors

$$2.38259e - 15, \quad 2.15741e - 15, \quad 6.57048e - 16, \quad 1.32035e - 15, \quad 5.19311e - 15,$$
$$3.31501e - 15, \quad 2.23614e - 16, \quad 4.50822e - 15, \quad 1.61329e - 14, \quad 2.38840e - 15,$$
$$8.90876e - 16, \quad 2.29554e - 15, \quad 2.29554e - 15,$$

and the componentwise backward errors (take $\mathcal{U}$ as the all-one tensor and $\mathcal{V}$ as the all-zero tensor)

$$1.82074e - 15, \quad 2.39923e - 15, \quad 3.62700e - 16, \quad 8.57883e - 16, \quad 3.10390e - 15,$$
$$2.12358e - 15, \quad 1.05385e - 16, \quad 6.13387e - 15, \quad 1.42690e - 14, \quad 1.06205e - 15,$$
$$4.79252e - 16, \quad 1.30099e - 15, \quad 1.30099e - 15.$$

From the backward errors we can conclude that these eigenvalues and eigenvectors are all computed accurately and the least accurately computed eigenvalue is the minimal real eigenvalue $\lambda = -1.09531$.

We compute the sign-complex spectral radius of the transformed tensor pair $\{\widetilde{\mathcal{A}}, \widetilde{\mathcal{B}}\}$ by solving the optimization problem (2.15) by the function `fminimax` in **Matlab**. The computational result is $\rho^{\mathbb{C}}(\widetilde{\mathcal{A}}, \widetilde{\mathcal{B}}) = 21.983608$, which is attained by

$$\mathbf{x} = (0.5915, -0.7467, -0.3043)^{\top}.$$

From the previous computations, we know that the spectral radius $\rho(\widetilde{\mathcal{A}}, \widetilde{\mathcal{B}}) = 21.983566$, which is slightly smaller than $\rho^{\mathbb{C}}(\widetilde{\mathcal{A}}, \widetilde{\mathcal{B}})$ numerically. (Perhaps they are exactly the same in the absence of numerical errors.) In this example, the sign-complex spectral radius is a very good bound of the spectral radius of a tensor pair.

Finally, we focus on the real eigenvalues with real eigenvectors of $\{\mathcal{A}, \mathcal{B}\}$, that is, the Z-eigenvalues of $\mathcal{A}$. Since $\mathcal{A}$ is not positive definite (three of its Z-eigenvalues are negative), we can infer that there must be some real vector $\mathbf{x}$ such that $\mathcal{A}\mathbf{x}^4 = 0$ by Lemma 2.11. Therefore the Crawford number of the tensor pair $\{\mathcal{A}, \mathcal{B}\}$ is exactly $c_r(\mathcal{A}, \mathcal{B}) = 1$. Randomly construct a perturbation tensor $\mathcal{E}$ whose L-norm is $1e - 6$. Compare the extremal eigenvalues of $\{\mathcal{A}, \mathcal{B}\}$ and $\{\mathcal{A} + \mathcal{E}, \mathcal{B}\}$,

$$\text{chord}(\lambda_{\max}, \widetilde{\lambda}_{\max}) = 3.6447e - 7, \quad \text{chord}(\lambda_{\min}, \widetilde{\lambda}_{\min}) = 5.8513e - 7.$$

Both are less than $\|\mathcal{E}\|_L/c_r(\mathcal{A}, \mathcal{B}) = 1e - 6$, which verifies the perturbation bound in Theorem 2.14. In fact, every Z-eigenvalue is perturbed within this bound, but we have not proven this result for the internal eigenvalues.

Part II

Hankel Tensors

Chapter 3

Fast Tensor-Vector Products

Hankel structures are widely employed in data analysis and signal processing. Not only Hankel matrices but also higher-order Hankel tensors arise frequently in disciplines such as exponential data fitting [13, 39, 93, 94, 89], frequency domain subspace identification [111], multidimensional seismic trace interpolation [118], and so on. As far as we know, the term "Hankel tensor" was first introduced by Luque and Thibon [80]. Boyer et al. [5, 11] discussed the higher-order singular value decompositions (HOSVD) of structured tensors, including symmetric tensors, Hankel tensors, and Toeplitz tensors in more detail. Moreover, Papy et al. employed Hankel-type tensors in exponential data fitting [13, 93, 94]. De Lathauwer [34] also concerned the "separation" of signals that can be modeled as sums of exponentials (or more generally, as exponential polynomials) by Hankel tensor approaches. As for the properties of Hankel tensors, Qi [102] recently investigated the spectral properties of Hankel tensor largely via the generating function. Song and Qi [112] investigated the spectral properties of Hilbert tensors, which are special Hankel tensors.

3.1 Hankel Tensors

An mth-order tensor $\mathcal{H} \in \mathbb{C}^{n_1 \times n_2 \times \cdots \times n_m}$ is called a *Hankel tensor* if

$$\mathcal{H}_{i_1 i_2 \ldots i_m} = \phi(i_1 + i_2 + \cdots + i_m)$$

for all $i_k = 0, 1, \ldots, n_k - 1$ $(k = 1, 2, \ldots, m)$. We call $\mathcal{H}$ a *square Hankel tensor* when $n_1 = n_2 = \cdots = n_m$. Note that the degree of freedom of a Hankel tensor is $d_{\mathcal{H}} := n_1 + n_2 + \cdots + n_m - m + 1$. Thus a vector $\mathbf{h}$ of length $d_{\mathcal{H}}$ called the *generating vector* of $\mathcal{H}$, defined by

$$h_k = \phi(k), \ k = 0, 1, \ldots, d_{\mathcal{H}} - 1$$

can completely determine the Hankel tensor $\mathcal{H}$, when the tensor size is fixed. Furthermore, when the entries of $\mathbf{h}$ are

$$h_k = \int_{-\infty}^{+\infty} t^k f(t) \mathrm{d}t,$$

Theory and Computation of Tensors.
http://dx.doi.org/10.1016/B978-0-12-803953-3.50003-4

then we call $f(t)$ the *generating function* of $\mathcal{H}$. The generating function of a square Hankel tensor is essential for studying its eigenvalues, positive semidefiniteness, and copositiveness [102].

Fast algorithms for Hankel or Toeplitz matrix-vector products involving fast Fourier transformations (FFT) are well known [17, 77, 84, 126]. However fast Hankel tensor computations are seldomly discussed. We propose an analogous fast algorithm for Hankel tensor-vector products, which has its applications to exponential data fitting.

3.2   Exponential Data Fitting

We begin with one of the sources of Hankel tensors and see where we need fast Hankel tensor-vector products. Exponential data fitting is very important in many applications in scientific computing and engineering, which represents the signals as a sum of exponentially damped sinusoids. The computations and applications of exponential data fitting have been studied by many, and readers can refer to [95, 96, 97].

Papy et al. [93, 94] introduced a higher-order tensor approach into exponential data fitting by connecting it with the Vandermonde decomposition of a Hankel tensor. As stated in [93], their algorithm is a higher-order variant of the Hankel total least squares (HTLS) method. HTLS is a modification of the famous ESPRIT algorithm [107, 65] by employing the total least squares (TLS [49]) instead of the least squares (LS), which enhances the robustness because the TLS is an errors-in-variables regression. Furthermore, Papy et al. concluded from numerical experiments that the Hankel tensor approach can perform better for some difficult situations than the classical one based on Hankel matrices, although there are no known results regarding how to choose the optimal size of the Hankel tensor.

In order to understand the necessity of fast algorithms for Hankel and block Hankel tensors, we sketch Papy's algorithm in this section and extend it to multidimensional exponential data fitting.

3.2.1   The One-Dimensional Case

Assume that we get a one-dimensional noiseless signal with N complex samples $\{x_n\}_{n=0}^{N-1}$, and this signal is modeled as a sum of K exponentially damped complex sinusoids, that is,

$$x_n = \sum_{k=1}^{K} a_k \exp(\iota\varphi_k) \exp\left((-\alpha_k + \iota\omega_k)n\Delta t\right),$$

where $\iota = \sqrt{-1}$, Δt is the sampling interval, while the amplitudes a_k, the phases φ_k, the damping factors α_k, and the pulsations ω_k are parameters of the model required to be estimated. The signal can also be expressed as

$$x_n = \sum_{k=1}^{K} c_k z_k^n,$$

where $c_k = a_k \exp(\imath\varphi_k)$ and $z_k = \exp\big((-\alpha_k + \imath\omega_k)\Delta t\big)$. Here c_k is called the k-th *complex amplitude* including the phase and z_k is called the k-th *pole* of the signal.

Part of the aim of exponential data fitting is to estimate the poles $\{z_k\}_{k=1}^{K}$ from the data $\{x_n\}_{n=0}^{N-1}$. After fixing the poles, we can obtain the complex amplitudes by solving a Vandermonde system.

Denote vector $\mathbf{x} = (x_0, x_1, \ldots, x_{N-1})^\top$. We first construct a Hankel tensor $\mathcal{H}$ of a fixed size $I_1 \times I_2 \times \cdots \times I_m$ with the generating vector $\mathbf{x}$. When $m = 3$, for instance, the Hankel tensor $\mathcal{H}$ is

(3.1)

$$
\mathcal{H}_{:,:,1} =
\begin{bmatrix}
x_0 & x_1 & \cdots & x_{I_2-2} & x_{I_2-1} \\
x_1 & \reflectbox{$\ddots$} & \reflectbox{$\ddots$} & \reflectbox{$\ddots$} & \vdots \\
\vdots & \reflectbox{$\ddots$} & \reflectbox{$\ddots$} & \reflectbox{$\ddots$} & x_{I_1+I_2-3} \\
x_{I_1-1} & x_{I_1} & \cdots & x_{I_1+I_2-3} & x_{I_1+I_2-2}
\end{bmatrix},
$$

$$
\mathcal{H}_{:,:,2} =
\begin{bmatrix}
x_1 & x_2 & \cdots & x_{I_2-1} & x_{I_2} \\
x_2 & \reflectbox{$\ddots$} & \reflectbox{$\ddots$} & \reflectbox{$\ddots$} & \vdots \\
\vdots & \reflectbox{$\ddots$} & \reflectbox{$\ddots$} & \reflectbox{$\ddots$} & x_{I_1+I_2-2} \\
x_{I_1} & x_{I_1+1} & \cdots & x_{I_1+I_2-2} & x_{I_1+I_2-1}
\end{bmatrix},
$$

$$
\vdots \qquad \vdots \qquad \vdots \qquad \vdots \qquad \vdots \qquad \vdots
$$

$$
\mathcal{H}_{:,:,I_3} =
\begin{bmatrix}
x_{I_3-1} & x_{I_3} & \cdots & x_{I_2+I_3-3} & x_{I_2+I_3-2} \\
x_{I_3} & \reflectbox{$\ddots$} & \reflectbox{$\ddots$} & \reflectbox{$\ddots$} & \vdots \\
\vdots & \reflectbox{$\ddots$} & \reflectbox{$\ddots$} & \reflectbox{$\ddots$} & x_{I_1+I_2+I_3-4} \\
x_{I_1+I_3-2} & x_{I_1+I_3-1} & \cdots & x_{I_1+I_2+I_3-4} & x_{I_1+I_2+I_3-3}
\end{bmatrix}.
$$

The order m can be chosen arbitrarily, and the sizes I_p of each dimension should be no less than K and satisfy $I_1 + I_2 + \cdots + I_m - m + 1 = N$. Papy et al. [93] verified that the Vandermonde decomposition of $\mathcal{H}$ is

$$
\mathcal{H} = \mathcal{C} \times_1 Z_1^\top \times_2 Z_2^\top \cdots \times_m Z_m^\top,
$$

where $\mathcal{C}$ is a diagonal tensor with diagonal entries $\{c_k\}_{k=1}^{K}$, and each Z_p is a Vandermonde matrix [49]

$$
Z_p^\top =
\begin{bmatrix}
1 & z_1 & z_1^2 & \cdots & z_1^{I_p-1} \\
1 & z_2 & z_2^2 & \cdots & z_2^{I_p-1} \\
\vdots & \vdots & \vdots & \vdots & \vdots \\
1 & z_K & z_K^2 & \cdots & z_K^{I_p-1}
\end{bmatrix}.
$$

Therefore the problem will be solved if we obtain the Vandermonde decomposition of this Hankel tensor $\mathcal{H}$.

In [93] the Vandermonde matrices are estimated by applying the TLS to the factor matrices in the higher-order singular value decomposition (HOSVD, [35, 66]) of the best rank-$(R, R, \ldots, R)$ approximation [36, 66] of $\mathcal{H}$. If K is known, then take $R = K$. Otherwise, when it is unknown, R should be chosen to be much larger than an estimate of K. Therefore the HOSVD of the best low rank approximation of a Hankel tensor is the dominant part of the computation in exponential data fitting.

De Lathauwer et al. [36] proposed an effective algorithm called the higher-order orthogonal iterations (HOOI) for this purpose. There are other algorithms with faster convergence such as [37, 45] proposed, and one can refer to [66] for more details. Nevertheless, HOOI is still very popular because it is so simple and effective, as in [93].

The original HOOI algorithm for general tensors is displayed as follows, and the output $\mathcal{S} \times_1 U_1^\top \times_2 U_2^\top \cdots \times_m U_m^\top$ is the best rank-$(R_1, R_2, \ldots, R_m)$ approximation [34, 36, 37, 45] of $\mathcal{A}$.

Algorithm 3.1. *HOOI for the best rank-$(R_1, R_2, \ldots, R_m)$ approximation of $\mathcal{A} \in \mathbb{C}^{I_1 \times I_2 \times \cdots \times I_m}$.*

> Initialize $U_p \in \mathbb{C}^{I_p \times R_p}$ $(p = 1, 2, \ldots, m)$ by the HOSVD of $\mathcal{A}$
> Repeat
> > for $p = 1 : m$
> > > $U_p \leftarrow R_p$ leading left singular vectors of
> > > $$\text{Unfold}_p(\mathcal{A} \times_1 \bar{U}_1 \cdots \widehat{\times_p \bar{U}_p} \cdots \times_m \bar{U}_m)$$
> > end
> Until convergence
> $\mathcal{S} = \mathcal{A} \times_1 \bar{U}_1 \times_2 \bar{U}_2 \cdots \times_m \bar{U}_m$.

Here $\text{Unfold}_p(\cdot)$ denotes the mode-p unfolding of a tensor as defined in Chapter 1, and $\mathcal{A} \times_1 \bar{U}_1 \cdots \widehat{\times_p \bar{U}_p} \cdots \times_m \bar{U}_m$ means that we skip the p-th item. There are plenty of tensor-matrix products in the above algorithm, which can be implemented using tensor-vector products. For instance, the tensor-matrix product

$$(\mathcal{A} \times_2 \bar{U}_2 \cdots \times_m \bar{U}_m)_{:,i_2,\ldots,i_m} = \mathcal{A} \times_2 (\bar{U}_2)_{:,i_2} \cdots \times_m (\bar{U}_m)_{:,i_m},$$

and others are the same. Therefore if all the Hankel tensor-vector products can be computed fast, then the HOOI algorithm for exponential data fitting will be efficient.

Papy et al. also studied the multi-channel and the decimative cases of exponential data fitting in [93, 94] using tensors. The tensors arise from these cases are not exactly Hankel tensors, but they have some Hankel structures. A tensor is called a *partially Hankel tensor*, if the lower-order subtensors are all Hankel tensors when some indices are fixed. For instance, the tensor $\mathcal{H}$ from the multi-channel or decimative exponential data fitting is 3rd-order, and $\mathcal{H}(:,:,k)$ are Hankel tensors for all k, so we call $\mathcal{H}$ a 3rd-order $(1, 2)$-Hankel tensor. The HOOI algorithm will be applied to some partially Hankel tensors. Hence the fast tensor-vector products for partially Hankel tensors have to be considered.

3.2.2   The Multidimensional Case

Papy's method [93, 94] can be extended to multidimensional exponential data fitting, which involves block tensors. Similar to block matrices [17, 84], a *block tensor* has entries or blocks being tensors. The size of each block is called the *level-1 size* of the block tensor, and the size of the block-entry tensor is called the *level-2 size*. Furthermore, a *level-d block tensor* is a tensor whose entries are level-$(d - 1)$ block tensors.

We take the two-dimensional exponential data fitting [106, 123] as an example to illustrate our block tensor approach. Assume that there is a 2D noiseless signal with $N_1 \times N_2$ complex samples $\{x_{n_1 n_2}\}_{\substack{n_1 = 0,1,\ldots,N_1-1 \\ n_2 = 0,1,\ldots,N_2-1}}$ which is modeled as a sum of K exponential terms

$$x_{n_1 n_2} = \sum_{k=1}^{K} a_k \exp(\imath \varphi_k) \exp\left((-\alpha_{1,k} + \imath \omega_{1,k})n_1 \Delta t_1 + (-\alpha_{2,k} + \imath \omega_{2,k})n_2 \Delta t_2\right),$$

where the parameters are the same as those of 1D signals. Also, this 2D signal can be rewritten into a compact form

$$x_{n_1 n_2} = \sum_{k=1}^{K} c_k z_{1,k}^{n_1} z_{2,k}^{n_2}.$$

Our aim is still to estimate the poles $\{z_{1,k}\}_{k=1}^{K}$ and $\{z_{2,k}\}_{k=1}^{K}$ of the signal from the samples. We shall see shortly that the extended Papy's algorithm can also be regarded as a modified version of the 2D ESPRIT method [106].

Denote $X = (x_{n_1 n_2})_{N_1 \times N_2}$. Then we map the data X into a block Hankel tensor with Hankel blocks (BHHB tensor) $\mathcal{H}$ of level-1 size $I_1 \times I_2 \times \cdots \times I_m$ and level-2 size $J_1 \times J_2 \times \cdots \times J_m$. The sizes I_p and J_p of each dimension should be no less than K and satisfy that $I_1 + I_2 + \cdots + I_m - m + 1 = N_1$ and $J_1 + J_2 + \cdots + J_m - m + 1 = N_2$. First, we construct the Hankel tensors $\mathcal{H}_j$ of size $I_1 \times I_2 \times \cdots \times I_m$ with the generating vectors $X(:,j)$ for $j = 0, 1, \ldots, N_2-1$ as shown in (3.1). Then, in the block sense, we construct the block Hankel tensors $\mathcal{H}$ of size $J_1 \times J_2 \times \cdots \times J_m$ with the block generating vectors $[\mathcal{H}_0, \mathcal{H}_1, \ldots, \mathcal{H}_{N_2-1}]^\top$. For instance, when $m = 3$ the slices in block sense of BHHB tensor $\mathcal{H}$ are

$$\mathcal{H}_{:,:,1}^{(b)} = \begin{bmatrix} \mathcal{H}_0 & \mathcal{H}_1 & \cdots & \mathcal{H}_{J_2-2} & \mathcal{H}_{J_2-1} \\ \mathcal{H}_1 & \ddots & \ddots & \ddots & \vdots \\ \vdots & \ddots & \ddots & \ddots & \mathcal{H}_{J_1+J_2-3} \\ \mathcal{H}_{J_1-1} & \mathcal{H}_{J_1} & \cdots & \mathcal{H}_{J_1+J_2-3} & \mathcal{H}_{J_1+J_2-2} \end{bmatrix},$$

(3.2)
$$\mathcal{H}_{:,:,2}^{(b)} = \begin{bmatrix} \mathcal{H}_1 & \mathcal{H}_2 & \cdots & \mathcal{H}_{J_2-1} & \mathcal{H}_{J_2} \\ \mathcal{H}_2 & \ddots & \ddots & \ddots & \vdots \\ \vdots & \ddots & \ddots & \ddots & \mathcal{H}_{J_1+J_2-2} \\ \mathcal{H}_{J_1} & \mathcal{H}_{J_1+1} & \cdots & \mathcal{H}_{J_1+J_2-2} & \mathcal{H}_{J_1+J_2-1} \end{bmatrix},$$

$$\vdots$$

$$\mathcal{H}_{:,:,J_3}^{(b)} = \begin{bmatrix} \mathcal{H}_{J_3-1} & \mathcal{H}_{J_3} & \cdots & \mathcal{H}_{J_2+J_3-3} & \mathcal{H}_{J_2+J_3-2} \\ \mathcal{H}_{J_3} & \ddots & \ddots & \ddots & \vdots \\ \vdots & \ddots & \ddots & \ddots & \mathcal{H}_{J_1+J_2+J_3-4} \\ \mathcal{H}_{J_1+J_3-2} & \mathcal{H}_{J_1+J_3-1} & \cdots & \mathcal{H}_{J_1+J_2+J_3-4} & \mathcal{H}_{J_1+J_2+J_3-3} \end{bmatrix}.$$

Then the BHHB tensor $\mathcal{H}$ has the level-2 Vandermonde decomposition

$$\mathcal{H} = \mathcal{C} \times_1 \left(Z_{2,1} \underset{\mathrm{KR}}{\otimes} Z_{1,1}\right)^\top \times_2 \left(Z_{2,2} \underset{\mathrm{KR}}{\otimes} Z_{1,2}\right)^\top \cdots \times_m \left(Z_{2,m} \underset{\mathrm{KR}}{\otimes} Z_{1,m}\right)^\top,$$

where $\mathcal{C}$ is a diagonal tensor with diagonal entries $\{c_k\}_{k=1}^K$, each $Z_{1,p}$ or $Z_{2,p}$ is a Vandermonde matrix

$$Z_{1,p}^\top = \begin{bmatrix} 1 & z_{1,1} & z_{1,1}^2 & \cdots & z_{1,1}^{I_p-1} \\ 1 & z_{1,2} & z_{1,2}^2 & \cdots & z_{1,2}^{I_p-1} \\ \vdots & \vdots & \vdots & \vdots & \vdots \\ 1 & z_{1,K} & z_{1,K}^2 & \cdots & z_{1,K}^{I_p-1} \end{bmatrix}, \quad Z_{2,p}^\top = \begin{bmatrix} 1 & z_{2,1} & z_{2,1}^2 & \cdots & z_{2,1}^{J_p-1} \\ 1 & z_{2,2} & z_{2,2}^2 & \cdots & z_{2,2}^{J_p-1} \\ \vdots & \vdots & \vdots & \vdots & \vdots \\ 1 & z_{2,K} & z_{2,K}^2 & \cdots & z_{2,K}^{J_p-1} \end{bmatrix},$$

and the notation $\underset{\mathrm{KR}}{\otimes}$ denotes the *Khatri-Rao product* [49, Chapter 12.3] of two matrices with the same column sizes

$$[\mathbf{a}_1, \mathbf{a}_2, \ldots, \mathbf{a}_n] \underset{\mathrm{KR}}{\otimes} [\mathbf{b}_1, \mathbf{b}_2, \ldots, \mathbf{b}_n] = [\mathbf{a}_1 \otimes \mathbf{b}_1, \mathbf{a}_2 \otimes \mathbf{b}_2, \ldots, \mathbf{a}_n \otimes \mathbf{b}_n].$$

Therefore our aim is achieved if we obtain the level-2 Vandermonde decomposition of this BHHB tensor $\mathcal{H}$.

We can use HOOI as well to compute the best rank-$(K, K, \ldots, K)$ approximation of the BHHB tensor

$$\mathcal{H} = \mathcal{S} \times_1 U_1^\top \times_2 U_2^\top \cdots \times_m U_m^\top,$$

where $\mathcal{S} \in \mathbb{C}^{K \times K \times \cdots \times K}$ is the core tensor and each $U_p \in \mathbb{C}^{(I_p J_p) \times K}$ has orthogonal columns. Then U_p and $Z_{2,p} \underset{\mathrm{KR}}{\otimes} Z_{1,p}$ have the common column space, that is, there is a nonsingular matrix T such that

$$Z_{2,p} \underset{\mathrm{KR}}{\otimes} Z_{1,p} = U_p T.$$

Denote

$$A^{1\uparrow} = \left[A_{0:I-2,:}^\top, A_{I,2I-2,:}^\top, \ldots, A_{(J-1)I:JI-2,:}^\top \right]^\top,$$
$$A^{1\downarrow} = \left[A_{1:I-1,:}^\top, A_{I+1,2I-1,:}^\top, \ldots, A_{(J-1)I+1:JI-1,:}^\top \right]^\top,$$
$$A^{2\uparrow} = A_{0:(J-1)I-1,:},$$
$$A^{2\downarrow} = A_{I:JI-1,:},$$

for matrix $A \in \mathbb{C}^{(IJ) \times K}$. It is easy to verify that

$$\left(Z_{2,p} \underset{\mathrm{KR}}{\otimes} Z_{1,p} \right)^{1\uparrow} D_1 = \left(Z_{2,p} \underset{\mathrm{KR}}{\otimes} Z_{1,p} \right)^{1\downarrow}, \quad \left(Z_{2,p} \underset{\mathrm{KR}}{\otimes} Z_{1,p} \right)^{2\uparrow} D_2 = \left(Z_{2,p} \underset{\mathrm{KR}}{\otimes} Z_{1,p} \right)^{2\downarrow},$$

where D_1 is a diagonal matrix with diagonal entries $\{z_{1,k}\}_{k=1}^K$ and D_2 is a diagonal matrix with diagonal entries $\{z_{2,k}\}_{k=1}^K$. Then we have

$$U_p^{1\uparrow} (T D_1 T^{-1}) = U_p^{1\downarrow}, \quad U_p^{2\uparrow} (T D_2 T^{-1}) = U_p^{2\downarrow}.$$

Therefore if two matrices W_1 and W_2 satisfy

$$U_p^{1\uparrow} W_1 = U_p^{1\downarrow}, \quad U_p^{2\uparrow} W_2 = U_p^{2\downarrow},$$

then they share the same eigenvalues with D_1 and D_2, respectively. Equivalently, the eigenvalues of W_1 and W_2 are exactly the poles of the first and second dimension, respectively. Furthermore, we may choose the TLS as in [93] for solving the above two equations, since the noise on both sides should be taken into consideration.

Unlike the 2D ESPRIT method, we obtain the poles of both dimensions by introducing only one BHHB tensor rather than constructing two related BHHB matrices. Hence the matrices W_1 and W_2 have the same eigenvectors. This information is useful for finding the assignment of the poles, and the eigenvalues of W_1 and W_2 with the same eigenvector are grouped into a pair.

Recall that Algorithm 3.1 seeks for BHHB tensors in 2D exponential data fitting. Thus the fast algorithm for BHHB tensor-vector products is also essential for this situation. Moreover, when we deal with the exponential data fitting problems of higher dimensions, that is, 3D or higher, higher-level block Hankel tensors will be naturally involved. Therefore it is also required to derive a unified fast algorithm for higher-level block Hankel tensor-vector products.

3.3 Anti-Circulant Tensors

The fast algorithm for Hankel tensor-vector products is based on a class of special Hankel tensors called *anti-circulant tensor*.

Circulant matrices [32] are famous, which constitute a special class of Toeplitz matrices [17, 84]. The first column entries of a circulant matrix shift down when moving right, as shown in the following 3-by-3 example

$$\begin{bmatrix} c_0 & c_2 & c_1 \\ c_1 & c_0 & c_2 \\ c_2 & c_1 & c_0 \end{bmatrix}.$$

If the first column entries of a matrix shift up when moving right, such as

$$\begin{bmatrix} c_0 & c_1 & c_2 \\ c_1 & c_2 & c_0 \\ c_2 & c_0 & c_1 \end{bmatrix},$$

then it is a special Hankel matrix called an *anti-circulant matrix*, or a *left circulant* or *retrocirculant matrix* [32, Chapter 5]. Naturally, we generalize the anti-circulant matrix to the tensor case. A square Hankel tensor $\mathcal{C}$ of order m and dimension n is called an *anti-circulant tensor*, if its generating vector $\mathbf{h}$ satisfies

$$h_k = h_l, \quad \text{if } k \equiv l \,(\mathrm{mod}\, n).$$

Thus the generating vector is periodic

$$\mathbf{h} = (\underbrace{h_0, h_1, \ldots, h_{n-1}}_{\mathbf{c}^\top}, \underbrace{h_n, h_{n+1}, \ldots, h_{2n-1}}_{\mathbf{c}^\top}, \ldots, \underbrace{h_{(m-1)n}, \ldots, h_{m(n-1)}}_{\mathbf{c}(0:n-m)^\top})^\top.$$

Since the vector $\mathbf{c}$, which is exactly the "first" column $\mathcal{C}(:, 0, \cdots, 0)$, contains all the information about $\mathcal{C}$ and is more compact than the generating vector, we call it the *compressed generating vector* of the anti-circulant tensor. For instance, a $3 \times 3 \times 3$ anti-circulant tensor $\mathcal{C}$ is unfolded by mode-1 into

$$\mathrm{Unfold}_1(\mathcal{C}) = \left[\begin{array}{ccc|ccc|ccc} c_0 & c_1 & c_2 & c_1 & c_2 & c_0 & c_2 & c_0 & c_1 \\ c_1 & c_2 & c_0 & c_2 & c_0 & c_1 & c_0 & c_1 & c_2 \\ c_2 & c_0 & c_1 & c_0 & c_1 & c_2 & c_1 & c_2 & c_0 \end{array} \right],$$

and its compressed generating vector is $\mathbf{c} = [c_0, c_1, c_2]^\top$. Note that the degree of freedom of an anti-circulant tensor is always n, no matter how large its order m may be.

3.3.1   Diagonalization

One essential property of circulant matrices [32] is that each can be diagonalized by the Fourier matrix $F_n = \left(\exp(-\frac{2\pi\imath}{n} jk) \right)_{j,k=0,1,\ldots,n-1}$ ($\imath = \sqrt{-1}$). The Fourier matrix is exactly the Vandermonde matrix for the roots of unity, and it is unitary up to the normalization factor

$$F_n F_n^* = F_n^* F_n = n I_n,$$

where I_n is the identity matrix of $n \times n$ and F_n^* is the conjugate transpose of F_n. We will show that anti-circulant tensors also have a similar property, which brings much convenience for both analysis and computations.

In order to describe this property, we recall the definition and some properties of the mode-p tensor-matrix product. It should be pointed out that the tensor-matrix products in this paper are slightly different from some standard notations [35, 66]. Particularly, when $\mathcal{A}$ is a matrix, the mode-1 and mode-2 products can be written as

$$\mathcal{A} \times_1 M_1 \times_2 M_2 = M_1^\top \mathcal{A} M_2.$$

Notice that $M_1^\top \mathcal{A} M_2$ is totally different from $M_1^* \mathcal{A} M_2$! ($M_1^\top$ is the transpose of M_1.)

We state our main result on anti-circulant tensors.

Theorem 3.1. *A square tensor $\mathcal{C}$ of order m and dimension n is an anti-circulant tensor if and only if it can be diagonalized by the Fourier matrix F_n, that is,*

$$\mathcal{C} = \mathcal{D} F_n^m := \mathcal{D} \times_1 F_n \times_2 F_n \cdots \times_m F_n,$$

where $\mathcal{D}$ is a diagonal tensor and $\mathrm{diag}(\mathcal{D}) = \mathrm{ifft}(\mathbf{c})$. Here ifft is a Matlab-type symbol, an abbreviation of inverse fast Fourier transformation [120].

Proof. A direct check verifies that $\mathcal{D} F_n^m$ is anti-circulant. Thus we only need to prove that every anti-circulant tensor can be written in this form constructively.

First, assume that an anti-circulant tensor $\mathcal{C}$ could be written into $\mathcal{D} F_n^m$. How do we obtain the diagonal entries of $\mathcal{D}$ from $\mathcal{C}$? Since

$$\mathrm{diag}(\mathcal{D}) = \mathcal{D} \mathbf{1}^{m-1} = \frac{1}{n^m} \left(\mathcal{C}(F_n^*)^m \right) \mathbf{1}^{m-1} = \frac{1}{n^m} \bar{F}_n \left(\mathcal{C}(F_n^* \mathbf{1})^{m-1} \right)$$
$$= \frac{1}{n} \bar{F}_n (\mathcal{C} \mathbf{e}_0^{m-1}) = \frac{1}{n} \bar{F}_n \mathbf{c},$$

where $\mathbf{1} = [1, 1, \ldots, 1]^\top$, $\mathbf{e}_0 = [1, 0, \ldots, 0]^\top$, $\bar{F}_n$ is the conjugate of F_n, and $\mathbf{c}$ is the compressed generating vector of $\mathcal{C}$, we have

$$\mathrm{diag}(\mathcal{D}) = \mathrm{ifft}(\mathbf{c}).$$

Finally, it suffices to check that $\mathcal{C} = \mathcal{D} F^m$ with $\mathrm{diag}(\mathcal{D}) = \mathrm{ifft}(\mathbf{c})$ directly. Therefore every anti-circulant tensor can be diagonalized by the Fourier matrix of proper size. $\qquad\square$

From the expression $\mathcal{C} = \mathcal{D} F_n^m$, we have the corollary on the spectra of anti-circulant tensors.

Corollary 3.2. *An anti-circulant tensor $\mathcal{C}$ of order m and dimension n with the compressed generating vector $\mathbf{c}$ has a Z-/H-eigenvector $\frac{1}{\sqrt{n}}\mathbf{1}$, and the corresponding Z- and H-eigenvalues are $n^{\frac{m-2}{2}}\mathbf{1}^{\top}\mathbf{c}$ and $n^{m-2}\mathbf{1}^{\top}\mathbf{c}$, respectively. When n is even, it has another Z-eigenvector $\frac{1}{\sqrt{n}}\widetilde{\mathbf{1}}$, where $\widetilde{\mathbf{1}} = [1, -1, \ldots, 1, -1]^{\top}$, and the corresponding Z-eigenvalue is $n^{\frac{m-2}{2}}\widetilde{\mathbf{1}}^{\top}\mathbf{c}$; moreover, this is also an H-eigenvector if m is even, and the corresponding H-eigenvalue is $n^{m-2}\widetilde{\mathbf{1}}^{\top}\mathbf{c}$.*

Proof. It is easy to check that

$$\mathcal{C}\mathbf{1}^{m-1} = F_n^{\top}\big(\mathcal{D}(F_n\mathbf{1})^{m-1}\big) = n^{m-1}F_n^{\top}(\mathcal{D}\mathbf{e}_0^{m-1}) = n^{m-1}\mathcal{D}_{1,1,\ldots,1} \cdot F_n^{\top}\mathbf{e}_0$$
$$= n^{m-2}(\mathbf{e}_0^{\top}\bar{F}_n\mathbf{c})\mathbf{1} = (n^{m-2}\mathbf{1}^{\top}\mathbf{c})\mathbf{1}.$$

The proof of the rest part is similar. $\qquad\square$

3.3.2 Singular Values

Lim [76] defined the tensor singular values σ by

$$\begin{cases} \mathcal{A} \times_2 \mathbf{u}_2 \times_3 \mathbf{u}_3 \cdots \times_m \mathbf{u}_m = \varphi_{p_1}(\mathbf{u}_1) \cdot \sigma, \\ \mathcal{A} \times_1 \mathbf{u}_1 \times_3 \mathbf{u}_3 \cdots \times_m \mathbf{u}_m = \varphi_{p_2}(\mathbf{u}_2) \cdot \sigma, \\ \cdots \quad\quad \cdots \quad\quad \cdots \quad\quad \cdots \quad\quad \cdots \\ \mathcal{A} \times_1 \mathbf{u}_1 \times_2 \mathbf{u}_2 \cdots \times_{m-1} \mathbf{u}_{m-1} = \varphi_{p_m}(\mathbf{u}_m) \cdot \sigma, \end{cases}$$

where $\sigma \geq 0$ and $\mathbf{u}_l^{\top}\varphi_{p_l}(\mathbf{u}_l) = \|\mathbf{u}_l\|_{p_l}^{p_l} = 1$ for $l = 1, 2, \ldots, m$. When $p_1 = p_2 = \cdots = p_m = 2$, $\varphi_2(\mathbf{u}) = \bar{\mathbf{u}}$ and the singular values are unitarily invariant.

Consider the singular values of anti-circulant tensor $\mathcal{C} = \mathcal{D}F^m$. There exists a permutation matrix [49] P such that the diagonal entries of $\mathcal{D}P^m$ are arranged in descending order of their absolute values, then $\mathcal{C} = (\mathcal{D}P^m)\big((FP)^{\top}\big)^m$. Denote Λ a diagonal matrix satisfying $\Lambda_{kk}^m = \text{sgn}\big((\mathcal{D}P^m)_{kk}\big)$, where $\text{sgn}(\cdot)$ denotes the sign function

$$\text{sgn}(\xi) = \begin{cases} \xi/|\xi|, & \xi \neq 0, \\ 0, & \xi = 0. \end{cases}$$

Hence $\mathcal{C}$ can be rewritten as $\mathcal{C} = \widetilde{\mathcal{D}}(V^{\top})^m$, where $\widetilde{\mathcal{D}} = |\mathcal{D}P^m|$ is a nonnegative diagonal tensor with ordered diagonal entries and $V = FP\Lambda$ is unitary. If $\{\sigma; \mathbf{u}_1, \mathbf{u}_2, \ldots, \mathbf{u}_m\}$ is a singular tuple containing a singular value and the corresponding singular vectors of $\mathcal{C}$, then

$$\big\{\sigma; V^{\top}\mathbf{u}_1, V^{\top}\mathbf{u}_2, \ldots, V^{\top}\mathbf{u}_m\big\}$$

is a singular tuple of $\widetilde{\mathcal{D}}$, and vice versa. Therefore we need only to find the singular values and singular vectors of a diagonal tensor $\widetilde{\mathcal{D}}$. Let $d_1 \geq d_2 \geq \cdots \geq d_n \geq 0$ be the diagonal entries of $\widetilde{\mathcal{D}}$ and $\mathbf{w}_l = V^{\top}\mathbf{u}_l$ for $l = 1, 2, \ldots, m$. The singular value equations become

$$\begin{cases} d_k(\mathbf{w}_2)_k(\mathbf{w}_3)_k \ldots (\mathbf{w}_m)_k = (\bar{\mathbf{w}}_1)_k \cdot \sigma, \\ d_k(\mathbf{w}_1)_k(\mathbf{w}_3)_k \ldots (\mathbf{w}_m)_k = (\bar{\mathbf{w}}_2)_k \cdot \sigma, \\ \cdots \quad\quad \cdots \quad\quad \cdots \quad\quad \cdots \quad\quad \cdots \\ d_k(\mathbf{w}_1)_k(\mathbf{w}_2)_k \ldots (\mathbf{w}_{m-1})_k = (\bar{\mathbf{w}}_m)_k \cdot \sigma, \end{cases} \quad k = 1, 2, \ldots, n.$$

Consequently, we have for $k = 1, 2, \ldots, n$,

$$d_k(\mathbf{w}_1)_k(\mathbf{w}_2)_k \ldots (\mathbf{w}_m)_k = |(\mathbf{w}_1)_k|^2 \cdot \sigma = |(\mathbf{w}_2)_k|^2 \cdot \sigma = \cdots = |(\mathbf{w}_m)_k|^2 \cdot \sigma,$$

and $|\mathbf{w}_1| = |\mathbf{w}_2| = \cdots = |\mathbf{w}_m| := \mathbf{q} = [q_1, q_1, \ldots, q_n]^\top$ when $\sigma \neq 0$. Then $d_k q_k^{m-2} = \sigma$. Denote $K = \{k : q_k \neq 0\}$. Since $\mathbf{q}$ is normalized, we have $d_k > 0$ and $\sum_{k \in K} (\sigma/d_k)^{\frac{2}{m-2}} = 1$. Thus the singular value is

$$\sigma = \Big(\sum_{k \in K} d_k^{-\frac{2}{m-2}} \Big)^{-\frac{m-2}{2}},$$

and the singular vectors are determined by

$$q_k = \begin{cases} (\sigma/d_k)^{\frac{1}{m-2}}, & k \in K, \\ 0, & \text{otherwise,} \end{cases}$$

and

$$\mathrm{sgn}(\mathbf{w}_1)_k \mathrm{sgn}(\mathbf{w}_2)_k \ldots \mathrm{sgn}(\mathbf{w}_m)_k = 1.$$

Therefore, if $d_1 \geq \cdots \geq d_r > d_{r+1} = \cdots = d_n = 0$, then the anti-circulant tensor $\mathcal{C}$ has at most $2^r - 1$ nonzero singular values when $m > 2$, since the index set K can be chosen as an arbitrary subset of $\{1, 2, \ldots, n\}$. As for the zero singular values, the situation is a little more complicated. It is easy to verify that the above equation holds for some k if there are two of $\{(\mathbf{w}_1)_k, (\mathbf{w}_2)_k, \ldots, (\mathbf{w}_m)_k\}$ equal to zero. Furthermore, for $k = r + 1, r + 2, \ldots, n$, the k-th entries of $\mathbf{w}_l$ $(l = 1, 2, \ldots, m)$ can be chosen such that

$$\mathrm{sgn}(\mathbf{w}_1)_k \mathrm{sgn}(\mathbf{w}_2)_k \ldots \mathrm{sgn}(\mathbf{w}_m)_k = 1.$$

One can easily prove that the largest singular value of a nonnegative diagonal tensor is

$$d_1 = \max \sigma(\widetilde{\mathcal{D}})$$
$$= \max \big\{ |\widetilde{\mathcal{D}} \times_1 \mathbf{w}_1 \cdots \times_m \mathbf{w}_m| : \|\mathbf{w}_1\|_2 = \cdots = \|\mathbf{w}_m\|_2 = 1 \big\}.$$

Thus we also have that the largest singular value of an anti-circulant tensor

$$d_1 = \max \sigma(\mathcal{C})$$
$$= \max \big\{ |\mathcal{C} \times_1 \mathbf{u}_1 \cdots \times_m \mathbf{u}_m| : \|\mathbf{u}_1\|_2 = \cdots = \|\mathbf{u}_m\|_2 = 1 \big\},$$

and the maximum value can be attained when $\mathbf{u}_1 = \mathbf{u}_2 = \cdots = \mathbf{u}_m = \bar{V} \mathbf{e}_0$. Recall the definition of the tensor Z-eigenvalues [99]

$$\begin{cases} \mathcal{C} \mathbf{x}^{m-1} = \lambda \mathbf{x}, \\ \mathbf{x}^\top \mathbf{x} = 1, \end{cases}$$

where $\mathbf{x} \in \mathbb{R}^n$, then $\lambda = \mathcal{C} \mathbf{x}^m$.

Therefore the maximum absolute value of an anti-circulant tensor's Z-eigenvalues is bounded by the largest singular value, that is,

$$\rho_Z(\mathcal{C}) := \{|\lambda| : \lambda \text{ is a Z-eigenvalue of } \mathcal{C}\} \leq d_1.$$

Particularly, when the anti-circulant tensor $\mathcal{C}$ is further nonnegative, which is equivalent to its compressed generating vector $\mathbf{c}$ being nonnegative, it can be verified that

$$\mathrm{ifft}(\mathbf{c})_1 = \max_k |\mathrm{ifft}(\mathbf{c})_k|$$

where $\text{ifft}(\mathbf{c})_k$ denotes the k-th entry of $\text{ifft}(\mathbf{c})$.

Hence the singular vectors corresponding to the largest singular value are $\mathbf{u}_1 = \mathbf{u}_2 = \cdots = \mathbf{u}_m = \frac{1}{\sqrt{n}}\mathbf{1}$. Note that $\frac{1}{\sqrt{n}}\mathbf{1}$ is also a Z-eigenvector of $\mathcal{C}$ (see Corollary 3.2). Therefore the Z-spectral radius of a nonnegative anti-circulant tensor is exactly its largest singular value.

3.3.3 Block Tensors

Block structures arise in a variety of applications in scientific computing and engineering [64, 89]. We utilized block tensors to multidimensional data fitting in Section 3.2.2.

If a block tensor is a Hankel or an anti-circulant tensor with tensor entries, then we call it a *block Hankel* or a *block anti-circulant tensor*, respectively. Moreover, its generating vector $\mathbf{h}^{(b)}$ or compressed generating vector $\mathbf{c}^{(b)}$ in block sense is called the *block generating vector* or *block compressed generating vector*, respectively. For instance, the block-entry vector $[\mathcal{H}_0, \mathcal{H}_1, \ldots, \mathcal{H}_{N_2-1}]^\top$ is the block generating vector of $\mathcal{H}$ in Section 3.2.2. Recall the definition of Kronecker product [49]

$$A \otimes B = \begin{bmatrix} a_{11}B & a_{12}B & \cdots & a_{1q}B \\ a_{21}B & a_{22}B & \cdots & a_{2q}B \\ \vdots & \vdots & \ddots & \vdots \\ a_{p1}B & a_{p2}B & \cdots & a_{pq}B \end{bmatrix},$$

where A and B are two matrices of arbitrary sizes. Then it can be proved following Theorem 3.1 that a block anti-circulant tensor $\mathcal{C}$ can be block diagonalized by $F_N \otimes I$, that is,

$$\mathcal{C} = \mathcal{D}^{(b)}(F_N \otimes I)^m,$$

where $\mathcal{D}^{(b)}$ is a block diagonal tensor with diagonal blocks $\mathbf{c}^{(b)} \times_1 \left(\frac{1}{N}\bar{F}_N \otimes I\right)$ and $\bar{F}_N$ is the conjugate of F_N.

Furthermore, when the blocks of a block Hankel tensor are also Hankel tensors, we call it a *block Hankel tensor with Hankel blocks*, or BHHB tensor for short. Then its block generating vector can be reduced to a matrix, which is called the *generating matrix* H of a BHHB tensor:

$$H = [\mathbf{h}_0, \mathbf{h}_1, \ldots, \mathbf{h}_{N_1+\cdots+N_m-m}] \in \mathbb{C}^{(n_1+n_2+\cdots+n_m-m+1)\times(N_1+N_2+\cdots+N_m-m+1)},$$

where $\mathbf{h}_k$ is the generating vector of the k-th Hankel block in $\mathbf{h}^{(b)}$. For instance, the data matrix X is exactly the generating matrix of the BHHB tensor $\mathcal{H}$ in Section 3.2.2. Similarly, when the blocks of a block anti-circulant tensor are also anti-circulant tensors, we call it a *block anti-circulant tensor with anti-circulant blocks*, or BCCB tensor for short. Its *compressed generating matrix* C is defined by

$$C = [\mathbf{c}_0, \mathbf{c}_1, \ldots, \mathbf{c}_{N-1}] \in \mathbb{C}^{n \times N},$$

where $\mathbf{c}_k$ is the compressed generating vector of the k-th anti-circulant block in the block compressed generating vector $\mathbf{c}^{(b)}$. We can verify that a BCCB tensor $\mathcal{C}$ can be diagonalized by $F_N \otimes F_n$, that is,

$$\mathcal{C} = \mathcal{D}(F_N \otimes F_n)^m,$$

where $\mathcal{D}$ is a diagonal tensor with diagonal $\mathrm{diag}(\mathcal{D}) = \frac{1}{nN}\mathrm{vec}(\bar{F}_n C \bar{F}_N)$, which can be computed by 2D inverse fast Fourier transformation (IFFT2). Here $\mathrm{vec}(\cdot)$ denotes the vectorization operator [49].

We can even define higher-level block Hankel tensors. For instance, a block Hankel tensor with BHHB blocks is called a *level-3 block Hankel tensor*, and it is easily understood that it has a generating tensor of order 3. Generally, a block Hankel/anti-circulant tensor with level-$(k-1)$ block Hankel/anti-circulant blocks is called a *level-k block Hankel/anti-circulant tensor*. Furthermore, a level-k block anti-circulant tensor $\mathcal{C}$ can be diagonalized by $F_{n^{(k)}} \otimes F_{n^{(k-1)}} \otimes \cdots \otimes F_{n^{(1)}}$, that is,

$$\mathcal{C} = \mathcal{D}(F_{n^{(k)}} \otimes F_{n^{(k-1)}} \otimes \cdots \otimes F_{n^{(1)}})^m,$$

where $\mathcal{D}$ is a diagonal tensor that can be computed by multidimensional inverse fast Fourier transformation.

3.4 Fast Hankel Tensor-Vector Product

General tensor-vector products of high orders and large sizes without structures are expensive. For a square tensor $\mathcal{A}$ of order m and dimension n, the computational complexity of a tensor-vector product $\mathcal{A}\mathbf{x}^{m-1}$ or $\mathcal{A}\mathbf{x}^m$ is $\mathcal{O}(n^m)$. However, since Hankel and anti-circulant tensors have low degrees of freedom, it can be expected that a much faster algorithm for Hankel tensor-vector products exists. We focus on the following two tensor-vector products

$$\mathbf{y} = \mathcal{A} \times_2 \mathbf{x}_2 \cdots \times_m \mathbf{x}_m, \quad \alpha = \mathcal{A} \times_1 \mathbf{x}_1 \times_2 \mathbf{x}_2 \cdots \times_m \mathbf{x}_m,$$

which will be useful for applications.

The fast algorithm for anti-circulant tensor-vector products is easy to derive from Theorem 3.1. Let $\mathcal{C} = \mathcal{D}F_n^m$ be an anti-circulant tensor of order m and dimension n with the compressed generating vector $\mathbf{c}$. Then for vectors $\mathbf{x}_2, \mathbf{x}_3, \ldots, \mathbf{x}_m \in \mathbb{C}^n$, we have

$$\mathbf{y} = \mathcal{C} \times_2 \mathbf{x}_2 \cdots \times_m \mathbf{x}_m = F_n(\mathcal{D} \times_2 F_n\mathbf{x}_2 \cdots \times_m F_n\mathbf{x}_m).$$

Recall that $\mathrm{diag}(\mathcal{D}) = \mathrm{ifft}(\mathbf{c})$ and $F_n\mathbf{v} = \mathrm{fft}(\mathbf{v})$, where fft is a Matlab-type symbol for the fast Fourier transformation. Thus the fast procedure for the vector $\mathbf{y}$ is

$$\mathbf{y} = \mathrm{fft}\big(\mathrm{ifft}(\mathbf{c}).*\mathrm{fft}(\mathbf{x}_2).*\cdots.*\mathrm{fft}(\mathbf{x}_m)\big),$$

where $\mathbf{u}.*\mathbf{v}$ multiplies two vectors componentwise. Similarly, for vectors $\mathbf{x}_1, \mathbf{x}_2, \ldots, \mathbf{x}_m \in \mathbb{C}^n$, we have

$$\alpha = \mathcal{C} \times_1 \mathbf{x}_1 \times_2 \mathbf{x}_2 \cdots \times_m \mathbf{x}_m = \mathcal{D} \times_1 F_n\mathbf{x}_1 \times_2 F_n\mathbf{x}_2 \cdots \times_m F_n\mathbf{x}_m,$$

and the fast procedure for the scalar α is

$$\alpha = \mathrm{ifft}(\mathbf{c})^\top \big(\mathrm{fft}(\mathbf{x}_1).*\mathrm{fft}(\mathbf{x}_2).*\cdots.*\mathrm{fft}(\mathbf{x}_m)\big).$$

Since the computational complexity of either FFT or IFFT is $\mathcal{O}(n \log n)$, both types of anti-circulant tensor-vector products can be implemented with complexity $\mathcal{O}((m+1)n \log n)$, which is even much faster than the product of a general n-by-n matrix and a vector.

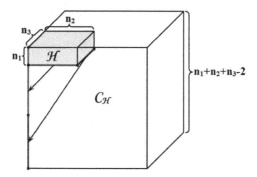

Figure 3.1: Embedding a Hankel tensor into an anti-circulant tensor.

For the fast algorithm for Hankel tensor-vector products, we embed a Hankel tensor into a larger anti-circulant tensor. Let $\mathcal{H} \in \mathbb{C}^{n_1 \times n_2 \times \cdots \times n_m}$ be a Hankel tensor with the generating vector $\mathbf{h}$. Denote $\mathcal{C}_{\mathcal{H}}$ the anti-circulant tensor of order m and dimension $d_{\mathcal{H}} = n_1 + n_2 + \cdots + n_m - m + 1$ with the compressed generating vector $\mathbf{h}$. The tensor $\mathcal{H}$ is in the upper left frontal corner of $\mathcal{C}_{\mathcal{H}}$ as shown in Fig. 3.1. Hence we have

$$\mathcal{C}_{\mathcal{H}} \times_2 \begin{bmatrix} \mathbf{x}_2 \\ \mathbf{0} \end{bmatrix} \cdots \times_m \begin{bmatrix} \mathbf{x}_m \\ \mathbf{0} \end{bmatrix} = \begin{bmatrix} \mathcal{H} \times_2 \mathbf{x}_2 \cdots \times_m \mathbf{x}_m \\ \dagger \end{bmatrix},$$

$$\mathcal{C}_{\mathcal{H}} \times_1 \begin{bmatrix} \mathbf{x}_1 \\ \mathbf{0} \end{bmatrix} \cdots \times_m \begin{bmatrix} \mathbf{x}_m \\ \mathbf{0} \end{bmatrix} = \mathcal{H} \times_1 \mathbf{x}_1 \cdots \times_m \mathbf{x}_m,$$

where $\dagger$ indicates the part of no interest. Therefore the Hankel tensor-vector products can be realized by multiplying a larger anti-circulant tensor with some augmented vectors. Therefore the fast procedure for $\mathbf{y} = \mathcal{H} \times_2 \mathbf{x}_2 \cdots \times_m \mathbf{x}_m$ is

$$\begin{cases} \widetilde{\mathbf{x}}_p = [\mathbf{x}_p^{\mathsf{T}}, \underbrace{0, 0, \ldots, 0}_{d_{\mathcal{H}} - n_p}]^{\mathsf{T}}, \ p = 2, 3, \ldots, m, \\ \widetilde{\mathbf{y}} = \mathrm{fft}\big(\mathrm{ifft}(\mathbf{h}). * \mathrm{fft}(\widetilde{\mathbf{x}}_2). * \cdots . * \mathrm{fft}(\widetilde{\mathbf{x}}_m)\big), \\ \mathbf{y} = \widetilde{\mathbf{y}}(0 : n_1 - 1), \end{cases}$$

and that for $\alpha = \mathcal{H} \times_1 \mathbf{x}_1 \times_2 \mathbf{x}_2 \cdots \times_m \mathbf{x}_m$ is

$$\begin{cases} \widetilde{\mathbf{x}}_p = [\mathbf{x}_p^{\mathsf{T}}, \underbrace{0, 0, \ldots, 0}_{d_{\mathcal{H}} - n_p}]^{\mathsf{T}}, \ p = 1, 2, \ldots, m, \\ \alpha = \mathrm{ifft}(\mathbf{h})^{\mathsf{T}} \big(\mathrm{fft}(\widetilde{\mathbf{x}}_1). * \mathrm{fft}(\widetilde{\mathbf{x}}_2). * \cdots . * \mathrm{fft}(\widetilde{\mathbf{x}}_m)\big). \end{cases}$$

Moreover, the computational complexity is $\mathcal{O}\big((m + 1)d_{\mathcal{H}} \log d_{\mathcal{H}}\big)$. When the Hankel tensor is a square tensor, the complexity is $\mathcal{O}(m^2 n \log mn)$, which is much smaller than the $\mathcal{O}(n^m)$ complexity of non-structured products.

Apart from the low computational complexity, our algorithm for Hankel tensor-vector products has two more advantages. One is that this scheme is optimal in the sense that it contains no redundant element. It is not required to form the Hankel tensor explicitly, and only the generating vector is needed. Another advantage is that our algorithm treats the tensor as an ensemble instead of multiplying the tensor by vectors mode by mode.

For the BCCB and BHHB cases, we also have fast algorithms for the tensor-vector products. Let $\mathcal{C}$ be a BCCB tensor of order m with the compressed generating matrix $C \in \mathbb{C}^{n \times N}$. Since $\mathcal{C}$ can be diagonalized by $F_N \otimes F_n$

$$\mathcal{C} = \mathcal{D}(F_N \otimes F_n)^m,$$

we have for vectors $\mathbf{x}_2, \mathbf{x}_3, \ldots, \mathbf{x}_m \in \mathbb{C}^{nN}$

$$\mathbf{y} = \mathcal{C} \times_2 \mathbf{x}_2 \cdots \times_m \mathbf{x}_m = (F_N \otimes F_n)\big(\mathcal{D} \times_2 (F_N \otimes F_n)\mathbf{x}_2 \cdots \times_m (F_N \otimes F_n)\mathbf{x}_m\big).$$

Recall the vectorization operator and its inverse operator

$$\text{vec}(A) = [A_{:,0}^\top, A_{:,1}^\top, \ldots, A_{:,N-1}^\top]^\top \in \mathbb{C}^{nN},$$
$$\text{vec}_{n,N}^{-1}(\mathbf{v}) = [\mathbf{v}_{0:n-1}, \mathbf{v}_{n:2n-1}, \ldots, \mathbf{v}_{(N-1)n:Nn-1}] \in \mathbb{C}^{n \times N},$$

for matrix $A \in \mathbb{C}^{n \times N}$ and vector $\mathbf{v} \in \mathbb{C}^{nN}$, and the relation

$$(B \otimes A)\mathbf{v} = \text{vec}\big(A \cdot \text{vec}_{n,N}^{-1}(\mathbf{v}) \cdot B^\top\big).$$

Therefore $(F_N \otimes F_n)\mathbf{x}_p = \text{vec}\big(F_n \cdot \text{vec}_{n,N}^{-1}(\mathbf{x}_p) \cdot F_N\big)$ can be computed by the 2D fast Fourier transformation (FFT2). Then the fast procedure for $\mathbf{y} = \mathcal{C} \times_2 \mathbf{x}_2 \cdots \times_m \mathbf{x}_m$ is

$$\begin{cases} X_p = \text{vec}_{n,N}^{-1}(\mathbf{x}_p), \ p = 2, 3, \ldots, m, \\ Y = \text{fft2}\big(\text{ifft2}(C). * \text{fft2}(X_2). * \cdots . * \text{fft2}(X_m)\big), \\ \mathbf{y} = \text{vec}(Y), \end{cases}$$

and that for $\alpha = \mathcal{C} \times_1 \mathbf{x}_1 \times_2 \mathbf{x}_2 \cdots \times_m \mathbf{x}_m$ is

$$\begin{cases} X_p = \text{vec}_{n,N}^{-1}(\mathbf{x}_p), \ p = 1, 2, \ldots, m, \\ \alpha = \big\langle \text{ifft2}(C), \text{fft2}(X_1). * \text{fft2}(X_2). * \cdots . * \text{fft2}(X_m)\big\rangle, \end{cases}$$

where $\langle A, B \rangle$ denotes

$$\langle A, B \rangle = \sum_{j,k} A_{jk} B_{jk}.$$

For a BHHB tensor $\mathcal{H}$ with the generating matrix H, we do the embedding twice. First, we embed each Hankel block into a larger anti-circulant block, and then we embed the block Hankel tensor with anti-circulant blocks into a BCCB tensor $\mathcal{C}_\mathcal{H}$ in block sense. Notice that the compressed generating matrix of $\mathcal{C}_\mathcal{H}$ is exactly the generating matrix of $\mathcal{H}$. Hence we have the fast procedure for $\mathbf{y} = \mathcal{H} \times_2 \mathbf{x}_2 \cdots \times_m \mathbf{x}_m$:

$$\begin{cases} \widetilde{X}_p = \begin{bmatrix} \text{vec}_{n_p,N_p}^{-1}(\mathbf{x}_p) & O \\ O & O \end{bmatrix} \Big\} n_1 + n_2 + \cdots + n_m - m + 1, \ p = 2, 3, \ldots, m, \\ \underbrace{\phantom{\begin{bmatrix} \text{vec} \end{bmatrix}}}_{N_1 + N_2 + \cdots + N_m - m + 1} \\ \widetilde{Y} = \text{fft2}\big(\text{ifft2}(H). * \text{fft2}(\widetilde{X}_2). * \cdots . * \text{fft2}(\widetilde{X}_m)\big), \\ \mathbf{y} = \text{vec}\big(\widetilde{Y}(0 : n_1 - 1, 0 : N_1 - 1)\big). \end{cases}$$

Sometimes in applications there is no need for the vectorization in the last line as $\widetilde{Y}$ and $\mathbf{y}$ contain the same information. We also have the fast procedure for

$\alpha = \mathcal{H} \times_1 \mathbf{x}_1 \times_2 \mathbf{x}_2 \cdots \times_m \mathbf{x}_m$:

$$
\begin{cases}
\widetilde{X}_p = \underbrace{\begin{bmatrix} \text{vec}_{n_p, N_p}^{-1}(\mathbf{x}_p) & O \\ O & O \end{bmatrix}}_{N_1 + N_2 + \cdots + N_m - m + 1} \Bigg\} n_1 + n_2 + \cdots + n_m - m + 1, \quad p = 1, 2, \ldots, m, \\[2em]
\alpha = \big\langle \text{ifft2}(H), \text{fft2}(\widetilde{X}_1). * \text{fft2}(\widetilde{X}_2). * \cdots . * \text{fft2}(\widetilde{X}_m) \big\rangle.
\end{cases}
$$

Similarly, we can also derive the fast algorithms for higher-level block Hankel tensor-vector products using the multidimensional FFT.

3.5 Numerical Examples

In this section, we will verify the effect of our fast algorithms for Hankel and block Hankel tensor-vector products by several numerical examples.

We first construct

- 3rd-order square Hankel tensors of size $n \times n \times n$ ($n = 10, 20, \ldots, 100$), and

- 3rd-order square BHHB tensors of level-1 size $n_1 \times n_1 \times n_1$ and level-2 size $n_2 \times n_2 \times n_2$ ($n_1, n_2 = 5, 6, \ldots, 12$).

Then we compute the tensor-vector products $\mathcal{H} \times_2 \mathbf{x}_2 \times_3 \mathbf{x}_3$ using

- our proposed fast algorithm based on FFT, and

- the non-structured algorithm based on the definition.

The average running times of 1000 products are shown in Fig. 3.2. From the results, we can see that the running time of our algorithm increases far more slowly than that of the non-structured algorithm just as predicted by the theoretical analysis. Note that the Hankel or BHHB tensors do not have to be formed explicitly in either algorithm.

Next, we shall apply our algorithm to the problems from exponential data fitting in order to show its efficiency, several of which are borrowed from [93, 94]. We do the experiments for both the one- and the two-dimensional cases:

- A one-dimensional signal is modeled as

$$
x_n = \exp\big((-0.01 + 2\pi i 0.20)n\big) + \exp\big((-0.02 + 2\pi i 0.22)n\big) + e_n,
$$

where e_n is a complex white Gaussian noise (WGN).

- A two-dimensional signal is modeled as

$$
\begin{aligned}
x_{n_1 n_2} = {} & \exp\big((-0.01 + 2\pi i 0.20)n_1\big) \cdot \exp\big((-0.02 + 2\pi i 0.18)n_2\big) \\
& + \exp\big((-0.02 + 2\pi i 0.22)n_1\big) \cdot \exp\big((-0.01 - 2\pi i 0.20)n_2\big) + e_{n_1 n_2},
\end{aligned}
$$

where $e_{n_1 n_2}$ is a two-dimensional complex white Gaussian noise.

The 3rd-order approach is chosen for both cases. We test the running times of the rank-$(2, 2, 2)$ approximation, since these signals both have two peaks. Moreover, we shall compute the HOSVDs of these Hankel and BHHB tensors,

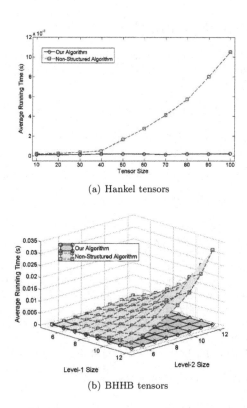

(a) Hankel tensors

(b) BHHB tensors

Figure 3.2: The average running time of tensor-vector products.

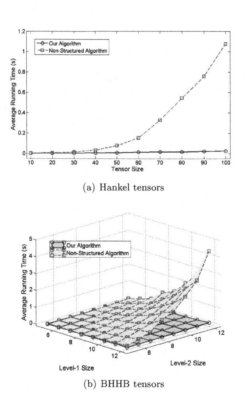

(a) Hankel tensors

(b) BHHB tensors

Figure 3.3: The average running time of HOOI.

which shows that Papy's algorithm [93, 94] and our extended multidimensional version can also work when the number of peaks is unknown.

Fig. 3.3 shows the comparison of these two algorithms' speeds. It provides a similar trend to the ones in Fig. 3.2, since the tensor-vector product plays a dominant role in the HOOI procedure. Therefore when the speed of tensor-vector products is largely increased by exploiting the Hankel or block Hankel structure, we can handle much larger problems than before.

Next, we fix the size of the Hankel and BHHB tensors. The Hankel tensor for the 1D exponential data fitting is of size $15 \times 15 \times 15$, and the BHHB tensor for 2D exponential data fitting is of level-1 size $5 \times 5 \times 5$ and level-2 size $6 \times 6 \times 6$. Assume that we do not know the number of peaks. Then we compute the HOSVD of the best rank-$(10, 10, 10)$ approximation

$$\mathcal{H} \approx \mathcal{S} \times_1 U_1^\top \times_2 U_2^\top \times_3 U_3^\top,$$

where the core tensor $\mathcal{S}$ is of size $10 \times 10 \times 10$. Fig. 3.4 displays the Frobenius norm of $\mathcal{S}(k, :, :)$ for $k = 1, 2, \ldots, 10$. We can see that the first two of them are apparently larger than the others. (The others should be zero when the signal is noiseless, but here we add a noise at 10^{-4}-level.) Thus we can conclude that the number of peaks is two. Furthermore, our fast algorithm enables us to accept a wild guess.

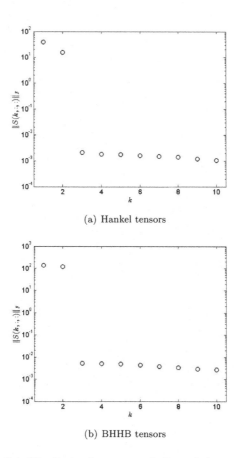

(a) Hankel tensors

(b) BHHB tensors

Figure 3.4: The Frobenius norms of slices of the core tensor.

Chapter 4

Inheritance Properties

We focus on real Hankel tensors in this chapter. It is well known that the positive semidefinite Hankel matrices are most related to the moment problems, and one can refer to [1, 47, 110]. In moment problems, necessary or sufficient conditions for the existence of a desired measure are given as the positive semidefiniteness of a series of Hankel matrices. We are also interested in the positive semidefiniteness of real Hankel tensors.

4.1 Inheritance Properties

Recall the concepts of Hankel tensors in Chapter 3. An mth-order *Hankel tensor* $\mathcal{H} \in \mathbb{C}^{n_1 \times n_2 \times \cdots \times n_m}$ is a multiway array whose entries are the function values of the sums of indices

$$\mathcal{H}_{i_1, i_2, \ldots, i_m} = h_{i_1 + i_2 + \cdots + i_m}, \ i_k = 0, 1, \ldots, n_k - 1, \ k = 1, 2, \ldots, m,$$

where the vector $\mathbf{h} = (h_i)_{i=0}^{n_1 + \cdots + n_m - m}$ is called the *generating vector* of this Hankel tensor $\mathcal{H}$ [39, 72, 102, 125]. The generating vector along with the size parameters determine the Hankel tensor. Furthermore, if the size parameters change, then the same vector can generate several Hankel tensors with different orders and sizes. What kinds of common properties will be shared for those Hankel tensors generated by the same vector? These common properties will be referred to inheritance properties in this chapter.

Each mth-order n-dimensional square tensor induces a degree-m multivariate polynomial of n variables:

$$p_{\mathcal{T}}(\mathbf{x}) := \mathcal{T}\mathbf{x}^m = \sum_{i_1=0}^{n-1} \cdots \sum_{i_m=0}^{n-1} \mathcal{T}_{i_1 i_2 \ldots i_m} x_{i_1} x_{i_2} \ldots x_{i_m}.$$

Suppose that m is even. If $p_{\mathcal{T}}(\mathbf{x})$ is always nonnegative (positive, nonpositive, or negative) for all nonzero real vectors $\mathbf{x}$, then the tensor $\mathcal{T}$ is called a *positive semidefinite (positive definite, negative semidefinite, or negative definite,* respectively) tensor [98]. If $p_{\mathcal{T}}(\mathbf{x})$ can be represented as a sum of squares, then the tensor $\mathcal{T}$ is called an *SOS tensor* [79]. Apparently, an SOS tensor must be positive semidefinite, but the converse is generally not true.

Theory and Computation of Tensors.
http://dx.doi.org/10.1016/B978-0-12-803953-3.50004-6

We organize this chapter in line with these spectral inheritance properties. However, other spectral inheritance properties, such as tensor decompositions (the Vandermonde and the SOS decompositions) and the convolution formula, will also be introduced as tools for investigating the spectral inheritance properties of Hankel tensors.

It is obvious that the positive definiteness is not well defined for odd-order tensors, since $\mathcal{T}(-\mathbf{x})^m = -\mathcal{T}\mathbf{x}^m$. Qi [98] proved that an even-order tensor is positive semidefinite if and only if it has no negative H-eigenvalues. Thus we shall use the property "no negative H-eigenvalue" instead of the positive semidefiniteness when studying the inheritance properties of odd-order Hankel tensors. The basic question about the inheritance of the positive semidefiniteness is:

- If a lower-order Hankel tensor has no negative H-eigenvalues, does a higher-order Hankel tensor with the same generating vector also possess no negative H-eigenvalues?

We will consider two cases:

- the lower order m is even and the higher order qm is a multiple of m, and

- the lower order is 2,

which provide the basic question with positive answers. Moreover, we guess that it is true when the lower order m is odd. However, we cannot prove it or find a counterexample, thus we leave it as conjecture.

In fact, Qi [102] showed an inheritance property of Hankel tensors. The generating vector of a Hankel tensor also generates a Hankel matrix, which is called the associated Hankel matrix of that Hankel tensor [102]. It was shown in [102] that if the Hankel tensor is of even order and its associated Hankel matrix is positive semidefinite, then the Hankel tensor is also positive semidefinite. In [102], a Hankel tensor is called a *strong Hankel tensor* if its associated Hankel matrix is positive semidefinite. Thus an even order strong Hankel tensor is positive semidefinite. It is actually the even-order case of the second situation. In this chapter, we show that the inheritance property holds in the second situation for odd orders.

The converse of the inheritance properties is not true. A simple case of the converse of the inheritance properties is as follows. Suppose a higher even order Hankel tensor is positive semidefinite. Is another higher-order Hankel matrix with the same generating vector positive semidefinite Hankel tensor also positive semidefinite? The answer is "no." Actually, if the answer were "yes," then all the even order positive semidefinite Hankel tensors would be strong Hankel tensors. In the literature, there are many examples of even order positive semidefinite Hankel tensors that are not strong Hankel tensors:

- the 4th-order 2-dimensional Hankel tensor [102] generated by

$$\left[1, 0, -\tfrac{1}{6}, 0, 1\right]^{\top},$$

- the 4th-order 4-dimensional Hankel tensor [26] generated by

$$\left[8, 0, 2, 0, 1, 0, 1, 0, 1, 0, 2, 0, 8\right]^{\top},$$

- the 6th-order 3-dimensional Hankel tensor [71] generated by

$$\left[h_0, 0, 0, 0, 0, 0, h_6, 0, 0, 0, 0, 0, h_{12}\right]^\top,$$

where $\sqrt{h_0 h_{12}} \geq \left(560 + 70\sqrt{70}\right) h_6 > 0$.

The following is a summary of the inheritance properties of Hankel tensors, studied in this chapter.

The first inheritance property of Hankel tensors is that if a lower-order Hankel tensor is positive semidefinite (positive definite, negative semidefinite, negative definite, or SOS), then its associated higher-order Hankel tensor with the same generating vector, where the higher order is a multiple of the lower order, is also positive semidefinite (positive definite, negative semidefinite, negative definite, or SOS, respectively). The inheritance property established in [102] can be regarded as a special case of this inheritance property. Furthermore, in this case, we show that the extremal H-eigenvalues of the higher-order tensor are bounded by the extremal H-eigenvalues of the lower-order tensor, multiplied with some constants.

In [72], it was proved that strong Hankel tensors are SOS tensors, but no concrete SOS decomposition was given. In this chapter, by using the inheritance property described above, we give a concrete sum-of-squares decomposition for a strong Hankel tensor.

The second inheritance property of Hankel tensors is an extension of the inheritance property established in [102] to the odd-order case. Normally, positive semidefiniteness and the SOS property are only well defined for even-order tensors. By [98], an even-order symmetric tensor is positive semidefinite if and only if it has no negative H-eigenvalues. In this chapter, we show that if the associated Hankel matrix of a Hankel tensor has no negative (nonpositive, positive, or nonnegative) eigenvalues, then the Hankel tensor has also no negative (nonpositive, positive, or nonnegative, respectively) H-eigenvalues. In this case, we show that the extremal H-eigenvalues of the Hankel tensor are bounded by the extremal eigenvalues of the associated Hankel matrix, multiplied with some constants.

Finally, we raise the third inheritance property of Hankel tensors as a conjecture.

4.2 The First Inheritance Property of Hankel Tensors

This section is devoted to the first inheritance property of Hankel tensors. We prove that if a lower-order Hankel tensor is positive semidefinite or SOS, then a Hankel tensor with the same generating vector and a higher multiple order is also positive semidefinite or SOS, respectively.

4.2.1 A Convolution Formula

We shall have a closer look at the nature of the Hankel structure first. In matrix theory, the multiplications of most structured matrices, such as Toeplitz, Hankel, Vandermonde, and Cauchy matrices, with vectors have their own analytic interpretations. Olshevsky and Shokrollahi [90] listed several important connections between fundamental analytic algorithms and structured matrix-vector multiplications. They claimed that there is a close relationship between Hankel matrices and discrete convolutions. Moreover, we will see shortly that this is also true for Hankel tensors.

We first introduce some basic facts about discrete convolutions. Given two vectors $\mathbf{u} \in \mathbb{C}^{n_1}$ and $\mathbf{v} \in \mathbb{C}^{n_2}$, their *convolution* $\mathbf{w} = \mathbf{u} * \mathbf{v} \in \mathbb{C}^{n_1+n_2-1}$ is a longer vector defined by

$$w_k = \sum_{j=0}^{k} u_j v_{k-j}, \quad k = 0, 1, \ldots, n_1 + n_2 - 2,$$

where $u_j = 0$ when $j \geq n_1$ and $v_j = 0$ when $j \geq n_2$. Denote $p_{\mathbf{u}}(\xi)$ and $p_{\mathbf{v}}(\xi)$ as the polynomials whose coefficients are $\mathbf{u}$ and $\mathbf{v}$, respectively

$$p_{\mathbf{u}}(\xi) = u_0 + u_1 \xi + \cdots + u_{n_1-1} \xi^{n_1-1}, \quad p_{\mathbf{v}}(\xi) = v_0 + v_1 \xi + \cdots + v_{n_2-1} \xi^{n_2-1}.$$

Then we can verify easily that $\mathbf{u} * \mathbf{v}$ consists of the coefficients of the product $p_{\mathbf{u}}(\xi) \cdot p_{\mathbf{v}}(\xi)$. Another important property of discrete convolutions is that

$$\mathbf{u} * \mathbf{v} = V^{-1}\left(V \begin{bmatrix} \mathbf{u} \\ \mathbf{0} \end{bmatrix} . * V \begin{bmatrix} \mathbf{v} \\ \mathbf{0} \end{bmatrix} \right)$$

for an arbitrary $(n_1 + n_2 - 1)$-by-$(n_1 + n_2 - 1)$ nonsingular Vandermonde matrix V, where $.*$ is a Matlab-type notation for multiplying two vectors component by component. In applications, the Vandermonde matrix is often taken as the Fourier matrices, since we have fast algorithms for discrete Fourier transforms [120]. Similarly, if there are more vectors $\mathbf{u}_1, \mathbf{u}_2, \ldots, \mathbf{u}_m$, then their convolution is equal to

$$(4.1) \qquad \mathbf{u}_1 * \mathbf{u}_2 * \cdots * \mathbf{u}_m = V^{-1}\left(V \begin{bmatrix} \mathbf{u}_1 \\ \mathbf{0} \end{bmatrix} . * V \begin{bmatrix} \mathbf{u}_2 \\ \mathbf{0} \end{bmatrix} . * \cdots . * V \begin{bmatrix} \mathbf{u}_m \\ \mathbf{0} \end{bmatrix} \right),$$

where V is a nonsingular Vandermonde matrix.

Recall the fast scheme for multiplying a Hankel tensor by vectors introduced in Chapter 3. The main approach is embedding a Hankel tensor into a larger anti-circulant tensor, which can be diagonalized by the Fourier matrices. We have proven that an mth-order N-dimensional anti-circulant tensor can be diagonalized by the N-by-N Fourier matrix

$$\mathcal{C} = \mathcal{D} \times_1 F_N \times_2 F_N \cdots \times_m F_N,$$

where $F_N = \left(\exp(\frac{2\pi \iota}{N} jk) \right)_{j,k=0}^{N-1}$ $(\iota = \sqrt{-1})$ is the N-by-N Fourier matrix, and $\mathcal{D}$ is a diagonal tensor with diagonal entries ifft$(\mathbf{c}) = F_N^{-1}\mathbf{c}$. Here, ifft is an abbreviation of inverse fast Fourier transform. Given m vectors $\mathbf{y}_1, \mathbf{y}_2, \ldots, \mathbf{y}_m \in \mathbb{C}^N$, we can calculate the anti-circulant tensor-vector product by

$$\mathcal{C} \times_1 \mathbf{y}_1 \times_2 \mathbf{y}_2 \cdots \times_m \mathbf{y}_m = (F_N^{-1}\mathbf{c})^\top (F_N \mathbf{y}_1 . * F_N \mathbf{y}_2 . * \cdots . * F_N \mathbf{y}_m),$$

where $F_N \mathbf{y}_k$ and $F_N^{-1}\mathbf{c}$ can be realized via fft and ifft, respectively.

Let $\mathcal{H}$ be an mth-order Hankel tensor of size $n_1 \times n_2 \times \cdots \times n_m$ and $\mathbf{h}$ be its generating vector. Taking the vector $\mathbf{h}$ as the compressed generating vector, we can form an anti-circulant tensor $\mathcal{C}_{\mathcal{H}}$ of order m and dimension $N = n_1 + \cdots + n_m - m + 1$. Interestingly, we find that the Hankel tensor $\mathcal{H}$ is exactly the first leading principal subtensor of $\mathcal{C}_{\mathcal{H}}$, that is, $\mathcal{H} = \mathcal{C}_{\mathcal{H}}(1 : n_1, 1 : n_2, \ldots, 1 : n_m)$. Hence the Hankel tensor-vector product $\mathcal{H} \times_1 \mathbf{x}_1 \times_2 \mathbf{x}_2 \cdots \times_m \mathbf{x}_m$ is equal to the anti-circulant tensor-vector product [39]

$$\mathcal{C}_{\mathcal{H}} \times_1 \begin{bmatrix} \mathbf{x}_1 \\ \mathbf{0} \end{bmatrix} \times_2 \begin{bmatrix} \mathbf{x}_2 \\ \mathbf{0} \end{bmatrix} \cdots \times_m \begin{bmatrix} \mathbf{x}_m \\ \mathbf{0} \end{bmatrix},$$

where $\mathbf{0}$ denotes an all-zero vector of appropriate size. Thus it can be computed via

$$\mathcal{H} \times_1 \mathbf{x}_1 \times_2 \mathbf{x}_2 \cdots \times_m \mathbf{x}_m = (F_N^{-1}\mathbf{h})^\top \left(F_N \begin{bmatrix} \mathbf{x}_1 \\ \mathbf{0} \end{bmatrix} . * F_N \begin{bmatrix} \mathbf{x}_2 \\ \mathbf{0} \end{bmatrix} . * \cdots . * F_N \begin{bmatrix} \mathbf{x}_m \\ \mathbf{0} \end{bmatrix} \right).$$

Particularly, when $\mathcal{H}$ is square and all the vectors are the same, that is, $n := n_1 = \cdots = n_m$ and $\mathbf{x} := \mathbf{x}_1 = \cdots = \mathbf{x}_m$, the homogeneous polynomial can be evaluated via

(4.2) $$\mathcal{H}\mathbf{x}^m = (F_N^{-1}\mathbf{h})^\top \left(F_N \begin{bmatrix} \mathbf{x} \\ \mathbf{0} \end{bmatrix} \right)^{[m]},$$

where $N = mn - m + 1$, and $\mathbf{v}^{[m]} = [v_1^m, v_2^m, \ldots, v_N^m]^\top$ stands for the componentwise m^{th} power of the vector $\mathbf{v}$. Moreover, this scheme has an analytic interpretation.

Comparing (4.2) and (4.1), we can write immediately that

(4.3) $$\mathcal{H}\mathbf{x}^m = \mathbf{h}^\top \underbrace{(\mathbf{x} * \mathbf{x} * \cdots * \mathbf{x})}_{m} =: \mathbf{h}^\top \mathbf{x}^{*m},$$

since $F_N = F_N^\top$. Employing this convolution formula for Hankel tensor-vector products, we can derive the inheritance properties of positive semidefiniteness and the SOS property of Hankel tensors from the lower order to the higher order.

4.2.2 Lower-Order Implies Higher-Order

Use $\mathcal{H}_m$ to denote an mth-order n-dimensional Hankel tensor with the generating vector $\mathbf{h} \in \mathbb{R}^{mn-m+1}$, where m is even and $n = qk - q + 1$ for some integers q and k. Then by the convolution formula (4.3), we have $\mathcal{H}_m \mathbf{x}^m = \mathbf{h}^\top \mathbf{x}^{*m}$ for an arbitrary vector $\mathbf{x} \in \mathbb{C}^n$. Assume that $\mathcal{H}_{qm}$ is a $(qm)^{\text{th}}$-order k-dimensional Hankel tensor that shares the same generating vector $\mathbf{h}$ with $\mathcal{H}_m$. Similarly, it holds that $\mathcal{H}_{qm}\mathbf{y}^{qm} = \mathbf{h}^\top \mathbf{y}^{*qm}$ for an arbitrary vector $\mathbf{y} \in \mathbb{C}^k$.

If $\mathcal{H}_m$ is positive semidefinite, then it is equivalent to $\mathcal{H}_m\mathbf{x}^m = \mathbf{h}^\top\mathbf{x}^{*m} \geq 0$ for all $\mathbf{x} \in \mathbb{R}^n$. Note that $\mathbf{y}^{*qm} = (\mathbf{y}^{*q})^{*m}$. Thus for an arbitrary vector $\mathbf{y} \in \mathbb{R}^k$, we have

$$\mathcal{H}_{qm}\mathbf{y}^{qm} = \mathbf{h}^\top\mathbf{y}^{*qm} = \mathbf{h}^\top(\mathbf{y}^{*q})^{*m} = \mathcal{H}_m(\mathbf{y}^{*q})^m \geq 0.$$

Therefore the higher-order but lower-dimensional Hankel tensor $\mathcal{H}_{qm}$ is also positive semidefinite. Furthermore, if $\mathcal{H}_m$ is positive definite, that is, $\mathcal{H}_m\mathbf{x}^m > 0$ for all nonzero vector $\mathbf{x} \in \mathbb{R}^n$, then $\mathcal{H}_{qm}$ is also positive definite. We may also derive the negative definite and negative semidefinite cases similarly.

If $\mathcal{H}_m$ is SOS, then there are some multivariate polynomials $p_1, p_2, \ldots, p_r$ such that for any $\mathbf{x} \in \mathbb{R}^n$

$$\mathcal{H}_m \mathbf{x}^m = \mathbf{h}^\top \mathbf{x}^{*m} = p_1(\mathbf{x})^2 + p_2(\mathbf{x})^2 + \cdots + p_r(\mathbf{x})^2.$$

Thus we have for any $\mathbf{y} \in \mathbb{R}^k$

$$(4.4) \qquad \mathcal{H}_{qm} \mathbf{y}^{qm} = \mathcal{H}_m (\mathbf{y}^{*q})^m = p_1(\mathbf{y}^{*q})^2 + p_2(\mathbf{y}^{*q})^2 + \cdots + p_r(\mathbf{y}^{*q})^2.$$

From the definition of discrete convolutions, we know that $\mathbf{y}^{*q}$ is also a multivariate polynomial about $\mathbf{y}$. Therefore the higher-order Hankel tensor $\mathcal{H}_{qm}$ is also SOS. Moreover, the SOS rank, that is, the minimum number of squares in the sum-of-squares representations [24], of $\mathcal{H}_{qm}$ is no larger than its SOS rank. Hence we summarize the inheritance properties of positive/negative semidefiniteness and the SOS property in the following theorem.

Theorem 4.1. *If an mth-order Hankel tensor is positive/negative (semi-)definite, then the $(qm)^{\text{th}}$-order Hankel tensor with positive integer q and the same generating vector inherits the same property. If an mth-order Hankel tensor is SOS, then the $(qm)^{\text{th}}$-order Hankel tensor with positive integer q and the same generating vector is also SOS with no larger SOS rank.*

Let $\mathcal{T}$ be an mth-order n-dimensional tensor and $\mathbf{x} \in \mathbb{C}^n$. Recall that $\mathcal{T}\mathbf{x}^{m-1}$ is a vector with $(\mathcal{T}\mathbf{x}^{m-1})_i = \sum_{i_2,\ldots,i_m=1}^n \mathcal{T}_{ii_2\ldots i_m} x_{i_2} \cdots x_{i_m}$. If there is a real scalar λ and a nonzero $\mathbf{x} \in \mathbb{R}^n$ such that $\mathcal{T}\mathbf{x}^{m-1} = \lambda \mathbf{x}^{[m-1]}$, where $\mathbf{x}^{[m-1]} := [x_1^{m-1}, x_2^{m-1}, \ldots, x_n^{m-1}]^\top$, then we call λ an *H-eigenvalue* of the tensor $\mathcal{T}$ and $\mathbf{x}$ a corresponding *H-eigenvector*. This concept was first introduced by Qi [98], and H-eigenvalues are shown to be essential for investigating tensors. By [98, Theorem 5], we know that an even-order symmetric tensor is positive (semi)definite if and only if all its H-eigenvalues are positive (nonnegative). Applying the convolution formula, we can further obtain a quantified result about the extremal H-eigenvalues of Hankel tensors.

Theorem 4.2. *Let $\mathcal{H}_m$ and $\mathcal{H}_{qm}$ be two Hankel tensors with the same generating vector of order m and qm, respectively, where m is even. Denote the minimal and the maximal H-eigenvalue of a tensor as $\lambda_{\min}(\cdot)$ and $\lambda_{\max}(\cdot)$, respectively. Then*

$$\lambda_{\min}(\mathcal{H}_{qm}) \geq \begin{cases} c_1 \cdot \lambda_{\min}(\mathcal{H}_m), & \text{if } \mathcal{H}_{qm} \text{ is positive semidefinite,} \\ c_2 \cdot \lambda_{\min}(\mathcal{H}_m), & \text{otherwise;} \end{cases}$$

and

$$\lambda_{\max}(\mathcal{H}_{qm}) \leq \begin{cases} c_1 \cdot \lambda_{\max}(\mathcal{H}_m), & \text{if } \mathcal{H}_{qm} \text{ is negative semidefinite,} \\ c_2 \cdot \lambda_{\max}(\mathcal{H}_m), & \text{otherwise,} \end{cases}$$

where $c_1 = \min_{\mathbf{y} \in \mathbb{R}^k} \|\mathbf{y}^{*q}\|_m^m / \|\mathbf{y}\|_{qm}^{qm}$ and $c_2 = \max_{\mathbf{y} \in \mathbb{R}^k} \|\mathbf{y}^{*q}\|_m^m / \|\mathbf{y}\|_{qm}^{qm}$ are positive constants depending on m, n, and q.

Proof. Since $\mathcal{H}_m$ and $\mathcal{H}_{qm}$ are even-order symmetric tensors, from [98, Theorem 5] we have

$$\lambda_{\min}(\mathcal{H}_m) = \min_{\mathbf{x} \in \mathbb{R}^n} \frac{\mathcal{H}_m \mathbf{x}^m}{\|\mathbf{x}\|_m^m}, \quad \lambda_{\min}(\mathcal{H}_{qm}) = \min_{\mathbf{y} \in \mathbb{R}^k} \frac{\mathcal{H}_{qm} \mathbf{y}^{qm}}{\|\mathbf{y}\|_{qm}^{qm}},$$

where $n = qk - q + 1$.

If $\mathcal{H}_{qm}$ is positive semidefinite, that is, $\mathcal{H}_{qm}\mathbf{y}^{qm} \geq 0$ for all $\mathbf{y} \in \mathbb{R}^k$, then we denote $c_1 = \min_{\mathbf{y} \in \mathbb{R}^k} \|\mathbf{y}^{*q}\|_m^m / \|\mathbf{y}\|_{qm}^{qm}$, which is a constant depending only on m, n, and q. Then by the convolution formula proposed above, we have

$$\lambda_{\min}(\mathcal{H}_{qm}) \geq c_1 \cdot \min_{\mathbf{y} \in \mathbb{R}^k} \frac{\mathcal{H}_{qm}\mathbf{y}^{qm}}{\|\mathbf{y}^{*q}\|_m^m} = c_1 \cdot \min_{\mathbf{y} \in \mathbb{R}^k} \frac{\mathcal{H}_m(\mathbf{y}^{*q})^m}{\|\mathbf{y}^{*q}\|_m^m}$$

$$\geq c_1 \cdot \min_{\mathbf{x} \in \mathbb{R}^n} \frac{\mathcal{H}_m\mathbf{x}^m}{\|\mathbf{x}\|_m^m} = c_1 \cdot \lambda_{\min}(\mathcal{H}_m).$$

If $\mathcal{H}_{qm}$ is not positive semidefinite, then we denote $c_2 = \max_{\mathbf{y} \in \mathbb{R}^k} \|\mathbf{y}^{*q}\|_m^m / \|\mathbf{y}\|_{qm}^{qm}$. Let $\widehat{\mathbf{y}}$ be a vector in $\mathbb{R}^k$ such that $\lambda_{\min}(\mathcal{H}_{qm}) = \mathcal{H}_{qm}\widehat{\mathbf{y}}^{qm} / \|\widehat{\mathbf{y}}\|_{qm}^{qm} < 0$. Then

$$\lambda_{\min}(\mathcal{H}_{qm}) \geq c_2 \cdot \frac{\mathcal{H}_{qm}\widehat{\mathbf{y}}^{qm}}{\|\widehat{\mathbf{y}}^{*q}\|_m^m} = c_2 \cdot \frac{\mathcal{H}_m(\widehat{\mathbf{y}}^{*q})^m}{\|\widehat{\mathbf{y}}^{*q}\|_m^m}$$

$$\geq c_2 \cdot \min_{\mathbf{x} \in \mathbb{R}^n} \frac{\mathcal{H}_m\mathbf{x}^m}{\|\mathbf{x}\|_m^m} = c_2 \cdot \lambda_{\min}(\mathcal{H}_m).$$

Thus we obtain a lower bound of the minimal H-eigenvalue of $\mathcal{H}_{qm}$, no matter whether this tensor is positive semidefinite or not. The proof of the upper bound of the maximal H-eigenvalue of $\mathcal{H}_{qm}$ is similar. $\qquad\square$

4.2.3 SOS Decomposition of Strong Hankel Tensors

When the lower order m in Theorem 4.1 equals 2, that is, the matrix case, the $(2q)^{\text{th}}$-order Hankel tensor sharing the same generating vector with this positive semidefinite Hankel matrix is called a *strong Hankel tensor*. We shall discuss strong Hankel tensors in detail in later sections. Now we focus on how to write out an SOS decomposition of a strong Hankel tensor following the formula (4.4). Li et al. [72] showed that even order strong Hankel tensors are SOS. However, their proof is not constructive, and no concrete SOS decomposition is given.

For an arbitrary Hankel matrix H generated by $\mathbf{h}$, we can compute its *Takagi factorization* efficiently by the algorithm proposed by Qiao et al. [14], where only the generating vector is required. The Takagi factorization can be written as $H = UDU^\top$, where $U = [\mathbf{u}_1, \mathbf{u}_2, \ldots, \mathbf{u}_r]$ is a column unitary matrix ($U^*U = I$) and $D = \text{diag}(d_1, d_2, \ldots, d_r)$. When the matrix is real, the Takagi factorization is exactly the singular value decomposition of the Hankel matrix H. Furthermore when H is positive semidefinite, the diagonal matrix D has nonnegative diagonal entries. Thus the polynomial $\mathbf{x}^\top H\mathbf{x}$ can be expressed as a sum of squares $p_1(\mathbf{x})^2 + p_2(\mathbf{x})^2 + \cdots + p_r(\mathbf{x})^2$, where

$$p_k(\mathbf{x}) = d_k^{1/2}\mathbf{u}_k^\top\mathbf{x}, \quad k = 1, 2, \ldots, r.$$

Following the formula (4.4), the $2q$-degree polynomial $\mathcal{H}_{2q}\mathbf{y}^{2q}$ can also be written as a sum of squares $q_1(\mathbf{y})^2 + q_2(\mathbf{y})^2 + \cdots + q_r(\mathbf{y})^2$, where

$$q_k(\mathbf{y}) = d_k^{1/2}\mathbf{u}_k^\top\mathbf{y}^{*q}, \quad k = 1, 2, \ldots, r.$$

Recall that any homogenous polynomial is associated with a symmetric tensor. An interesting observation is that the homogenous polynomial $q_k(\mathbf{y})$ is

---

Algorithm 4.1 An SOS decomposition of a strong Hankel tensor.

Input: The generating vector $\mathbf{h}$ of a strong Hankel tensor

Output: An SOS decomposition $q_1(\mathbf{y})^2 + q_2(\mathbf{y})^2 + \cdots + q_r(\mathbf{y})^2$ of this Hankel tensor

  1: Compute the Takagi factorization of the Hankel matrix generated by $\mathbf{h}$:
     $H = UDU^\top$

  2: $\mathbf{q}_k = d_k^{1/2}\mathbf{u}_k$ for $k = 1, 2, \ldots, r$

  3: Then $\mathbf{q}_k$ generates a qth-order Hankel tensor $\mathcal{Q}_k$ as the coefficient tensor of each term $q_k(\cdot)$ in the SOS decomposition for $k = 1, 2, \ldots, r$

---

associated with a qth-order Hankel tensor generated by $d_k^{1/2}\mathbf{u}_k$. Thus we determine an SOS decomposition of a strong Hankel tensor $\mathcal{H}_{2q}$ by r vectors $d_k^{1/2}\mathbf{u}_k$ ($k = 1, 2, \ldots, r$). We summarize the above procedure in Algorithm 4.1.

Example 4.1. *The first example is a 4th-order 3-dimensional Hankel tensor $\mathcal{H}$ generated by $[1, 0, 1, 0, 1, 0, 1, 0, 1]^\top \in \mathbb{R}^9$. The Takagi factorization of the Hankel matrix generated by the same vector is*

$$
\begin{bmatrix}
1 & 0 & 1 & 0 & 1 \\
0 & 1 & 0 & 1 & 0 \\
1 & 0 & 1 & 0 & 1 \\
0 & 1 & 0 & 1 & 0 \\
1 & 0 & 1 & 0 & 1
\end{bmatrix}
=
\begin{bmatrix}
\frac{1}{\sqrt{3}} & 0 \\
0 & \frac{1}{\sqrt{2}} \\
\frac{1}{\sqrt{3}} & 0 \\
0 & \frac{1}{\sqrt{2}} \\
\frac{1}{\sqrt{3}} & 0
\end{bmatrix}
\cdot
\begin{bmatrix}
3 & 0 \\
0 & 2
\end{bmatrix}
\cdot
\begin{bmatrix}
\frac{1}{\sqrt{3}} & 0 & \frac{1}{\sqrt{3}} & 0 & \frac{1}{\sqrt{3}} \\
0 & \frac{1}{\sqrt{2}} & 0 & \frac{1}{\sqrt{2}} & 0
\end{bmatrix}.
$$

Thus by Algorithm 4.1, an SOS decomposition of $\mathcal{H}\mathbf{y}^4$ is obtained:

$$
\left(
\begin{bmatrix} y_1 & y_2 & y_3 \end{bmatrix}
\cdot
\begin{bmatrix}
1 & 0 & 1 \\
0 & 1 & 0 \\
1 & 0 & 1
\end{bmatrix}
\cdot
\begin{bmatrix} y_1 \\ y_2 \\ y_3 \end{bmatrix}
\right)^2
+
\left(
\begin{bmatrix} y_1 & y_2 & y_3 \end{bmatrix}
\cdot
\begin{bmatrix}
0 & 1 & 0 \\
1 & 0 & 1 \\
0 & 1 & 0
\end{bmatrix}
\cdot
\begin{bmatrix} y_1 \\ y_2 \\ y_3 \end{bmatrix}
\right)^2
$$
$$
= (y_1^2 + y_2^2 + y_3^2 + 2y_1 y_3)^2 + (2y_1 y_2 + 2y_2 y_3)^2.
$$

However, the SOS decomposition is not unique, since $\mathcal{H}\mathbf{y}^4$ can also be written as $\frac{1}{2}(y_1 + y_2 + y_3)^4 + \frac{1}{2}(y_1 - y_2 + y_3)^4$.

4.3   The Second Inheritance Property of Hankel Tensors

In this section, we prove the second inheritance property of Hankel tensors, that is, if the associated Hankel matrix of a Hankel tensor has no negative (nonpositive, positive, or nonnegative) eigenvalues, then that Hankel tensor has no negative (nonpositive, positive, or nonnegative, respectively) H-eigenvalues. A basic tool to prove this is the augmented Vandermonde decomposition with positive coefficients.

4.3.1   Strong Hankel Tensors

Let $\mathcal{H}$ be an mth-order n-dimensional Hankel tensor generated by $\mathbf{h} \in \mathbb{R}^{mn-m+1}$. Then the square Hankel matrix H also generated by $\mathbf{h}$ is called the *associated*

Hankel matrix of the Hankel tensor $\mathcal{H}$. If the associated Hankel matrix of a Hankel tensor $\mathcal{H}$ is positive semidefinite, then we call this Hankel tensor $\mathcal{H}$ a *strong Hankel tensor* [102]. When m is even, we immediately know that an even-order strong Hankel tensor must be positive semidefinite by Theorem 4.1, which has been proved in [102, Theorem 3.1].

Qi [102] also introduced the *Vandermonde decomposition* of a Hankel tensor (see also [125])

$$
(4.5) \qquad \mathcal{H} = \sum_{k=1}^{r} \alpha_k \mathbf{v}_k^{\circ m},
$$

where $\mathbf{v}_k$ is in the Vandermonde form $\left[1, \xi_k, \xi_k^2, \ldots, \xi_k^{n-1}\right]^\top$, $\mathbf{v}^{\circ m} := \underbrace{\mathbf{v} \circ \mathbf{v} \circ \cdots \circ \mathbf{v}}_{m}$ is a rank-one tensor, and the outer product is defined by

$$
(\mathbf{v}_1 \circ \mathbf{v}_2 \circ \cdots \circ \mathbf{v}_m)_{i_1 i_2 \ldots i_m} = (\mathbf{v}_1)_{i_1} (\mathbf{v}_2)_{i_2} \cdots (\mathbf{v}_m)_{i_m}.
$$

The Vandermonde decomposition is equivalent to the factorization of the generating vector of $\mathcal{H}$

$$
\begin{bmatrix} h_0 \\ h_1 \\ h_2 \\ \vdots \\ h_r \end{bmatrix} = \begin{bmatrix} 1 & 1 & \cdots & 1 \\ \xi_1 & \xi_2 & \cdots & \xi_r \\ \xi_1^2 & \xi_2^2 & \cdots & \xi_r^2 \\ \vdots & \vdots & \vdots & \vdots \\ \xi_1^{mn-m} & \xi_2^{mn-m} & \cdots & \xi_r^{mn-m} \end{bmatrix} \cdot \begin{bmatrix} \alpha_1 \\ \alpha_2 \\ \vdots \\ \alpha_r \end{bmatrix}.
$$

Since the above Vandermonde matrix is nonsingular if and only if $r = mn-m+1$ and $\xi_1, \xi_2, \ldots, \xi_r$ are mutually distinct, every Hankel tensor must have such a Vandermonde decomposition with the number of terms r in (4.5) no larger than $mn - m + 1$. When the Hankel tensor is positive semidefinite, we desire that all the coefficients α_k are positive, so that each term in (4.5) is a rank-one positive semidefinite Hankel tensor when m is even. Moreover, a real square Hankel tensor $\mathcal{H}$ is called a *complete Hankel tensor*, if the coefficients α_k in one of its Vandermonde decompositions are all positive [102].

Nevertheless, the set of all complete Hankel tensors is not "complete." Li et al. show in [72, Corollary 1] that the mth-order n-dimensional complete Hankel tensor cone is not closed and its closure is the mth-order n-dimensional strong Hankel tensor cone. An obvious counterexample is $\mathbf{e}_n^{\circ m}$, where $\mathbf{e}_n = [0, \ldots, 0, 1]^\top$. Since all the Vandermonde vectors begin with unity and all α_k $(k = 1, 2, \ldots, r)$ are positive, the positive semidefinite Hankel tensor $\mathbf{e}_n^{\circ m}$ is not a complete Hankel tensor.

Fortunately, $\alpha \mathbf{e}_n^{\circ m}$ is the only kind of rank-one non-complete Hankel tensors.

Proposition 4.3. *If $\mathbf{v}^{\circ m}$ is a rank-one Hankel tensor, then*

$$
\mathbf{v} = \alpha[1, \xi, \xi^2, \ldots, \xi^{n-1}]^\top \quad or \quad \alpha[0, 0, \ldots, 0, 1]^\top.
$$

Proof. If $v_0 \neq 0$, then we can assume that $v_0 = 1$ without loss of generality. Denote $v_1 = \xi$. Then from the definition of Hankel tensors, we have $v_0^{m-2} v_1 v_{k-1} = v_0^{m-1} v_k$. Hence we can easily write $v_k = \xi^k$ for $k = 0, 1, \ldots, n-1$, that is, $\mathbf{v} = [1, \xi, \xi^2, \ldots, \xi^{n-1}]^\top$.

If $v_0 = 0$ and $k < n - 1$, then we have

(i) $v_{k-1}^{(m-1)/2} v_k v_{k+1}^{(m-1)/2} = v_k^m$ for m is odd;

(ii) $v_{k-1}^{m/2} v_{k+1}^{m/2} = v_k^m$ for m is even.

It implies that $u_k = 0$ for $k = 0, 1, \ldots, n-2$, so $\mathbf{v} = \alpha[0, 0, \ldots, 0, 1]^\top$.   □

We will soon show that if we add the term $\mathbf{e}_n^{\circ m}$ into the basis, then all the strong Hankel tensors can be decomposed into an *augmented Vandermonde decomposition*

$$\mathcal{H} = \sum_{k=1}^{r-1} \alpha_k \mathbf{v}_k^{\circ m} + \alpha_r \mathbf{e}_n^{\circ m}.$$

Note that $\frac{1}{\xi^{n-1}}[1, \xi, \xi^2, \ldots, \xi^{n-1}]^\top \to \mathbf{e}_n$ when $\xi \to \infty$. The cone of Hankel tensors with an augmented Vandermonde decomposition is actually the closure of the cone of complete Hankel tensors. When a Hankel tensor $\mathcal{H}$ has such an augmented Vandermonde decomposition, its associated Hankel matrix H also has a corresponding decomposition

$$H = \sum_{k=1}^{r-1} \alpha_k \widetilde{\mathbf{v}}_k^{\circ 2} + \alpha_r \mathbf{e}_{(n-1)m/2+1}^{\circ 2},$$

where $\widetilde{\mathbf{v}}_k = \left[1, \xi_k, \xi_k^2, \ldots, \xi_k^{(n-1)m/2}\right]^\top$ and $\mathbf{v}^{\circ 2}$ is exactly $\mathbf{v}\mathbf{v}^\top$, and vice versa. Therefore if a positive semidefinite Hankel tensor has an augmented Vandermonde decomposition with all positive coefficients, then it is a strong Hankel tensor, that is, its associated Hankel matrix must be positive semidefinite. Furthermore, when we obtain an augmented Vandermonde decomposition of its associated Hankel matrix, we can induce an augmented Vandermonde decomposition of the original Hankel tensor straightforwardly. Hence we begin with the positive semidefinite Hankel matrices.

4.3.2   A General Vandermonde Decomposition of Hankel Matrices

We shall introduce the algorithm for a general Vandermonde decomposition of an arbitrary Hankel matrix proposed by Boley et al. [12] in this subsection. Let us begin with a nonsingular Hankel matrix $H \in \mathbb{C}^{r \times r}$. After we solve the Yule-Walker equation [49, Chapter 4.7]:

$$\begin{bmatrix} h_0 & h_1 & h_2 & \cdots & h_{r-1} \\ h_1 & h_2 & h_3 & \cdots & h_r \\ \vdots & \vdots & \vdots & \ddots & \vdots \\ h_{r-2} & h_{r-1} & h_r & \cdots & h_{2r-3} \\ h_{r-1} & h_r & h_{r+1} & \cdots & h_{2r-2} \end{bmatrix} \cdot \begin{bmatrix} a_0 \\ a_1 \\ a_2 \\ \vdots \\ a_{r-1} \end{bmatrix} = \begin{bmatrix} h_r \\ h_{r+1} \\ \vdots \\ h_{2r-2} \\ \gamma \end{bmatrix},$$

we obtain an r term recurrence for $k = r, r+1, \ldots, 2r-2$

$$h_k = a_{r-1} h_{k-1} + a_{r-2} h_{k-2} + \cdots + a_0 h_{k-r}.$$

Denote C the companion matrix [49, Chapter 7.4.6] corresponding to the polynomial $p(\lambda) = \lambda^r - a_{r-1}\lambda^{r-1} - \cdots - a_0\lambda^0$, that is,

$$C = \begin{bmatrix} 0 & 1 & & & \\ & 0 & 1 & & \\ & & \ddots & \ddots & \\ & & & 0 & 1 \\ a_0 & a_1 & \cdots & a_{r-2} & a_{r-1} \end{bmatrix}.$$

Let the Jordan canonical form of C be $C = V^\top J V^{-\top}$, where $J = \mathrm{diag}(J_1, J_2, \ldots, J_s)$ and J_l is the $k_l \times k_l$ Jordan block corresponding to eigenvalue λ_l. Moreover, the nonsingular matrix V has the form

$$V = \left[\mathbf{v}, J^\top \mathbf{v}, (J^\top)^2 \mathbf{v}, \ldots, (J^\top)^{r-1}\mathbf{v}\right],$$

where $\mathbf{v} = [\mathbf{e}_{k_1,1}^\top, \mathbf{e}_{k_2,1}^\top, \ldots, \mathbf{e}_{k_s,1}^\top]^\top$ is a vector partitioned conformably with J and $\mathbf{e}_{k_l,1}$ is the first k_l-dimensional unit coordinate vector. This kind of V is often called a *confluent Vandermonde matrix* [12]. When the multiplicities of all the eigenvalues of C equal one, the matrix V is exactly a Vandermonde matrix.

Denote $\mathbf{h}_0$ as the first column of H and $\mathbf{w} = V^{-\top}\mathbf{h}_0$. There exists a unique block diagonal matrix $D = \mathrm{diag}(D_1, D_2, \ldots, D_s)$, which is also partitioned conformably with J, satisfying

$$D\mathbf{v} = \mathbf{w} \quad \text{and} \quad DJ^\top = JD.$$

Moreover, each block D_l is a k_l-by-k_l upper anti-triangular Hankel matrix. If we partition $\mathbf{w} = [\mathbf{w}_1, \mathbf{w}_2, \ldots, \mathbf{w}_s]^\top$ conformably with J, then the lth block is determined by

$$D_l = \begin{bmatrix} (\mathbf{w}_l)_1 & (\mathbf{w}_l)_2 & \cdots & (\mathbf{w}_l)_{k_l} \\ (\mathbf{w}_l)_2 & \cdots & (\mathbf{w}_l)_{k_l} & 0 \\ \vdots & \iddots & \iddots & \vdots \\ (\mathbf{w}_l)_{k_l} & 0 & \cdots & 0 \end{bmatrix}.$$

Finally, we obtain a general Vandermonde decomposition of the full-rank Hankel matrix

$$H = V^\top D V.$$

If the leading $r \times r$ principal submatrix, that is, $H(1 : r, 1 : r)$, of an $n \times n$ rank-r Hankel matrix H is nonsingular, then H admits the Vandermonde decomposition $H = (V_{r\times n})^\top D_{r\times r} V_{r\times n}$, which is induced by the decomposition of the leading $r \times r$ principal submatrix.

Nevertheless, this generalized Vandermonde decomposition is insufficient for establishing the positive definiteness of a real Hankel matrix, since the factors V and D could be complex even though H real. We shall modify this decomposition into a general real Vandermonde decomposition. Assume that two eigenvalues λ_1 and λ_2 of C form a pair of conjugate complex numbers. Then the corresponding parts in D and V are also conjugate, respectively. That is,

$$\begin{bmatrix} V_1^\top & V_2^\top \end{bmatrix} \cdot \begin{bmatrix} D_1 & \\ & D_2 \end{bmatrix} \cdot \begin{bmatrix} V_1 \\ V_2 \end{bmatrix} = \begin{bmatrix} V_1^\top & \bar{V}_1^\top \end{bmatrix} \cdot \begin{bmatrix} D_1 & \\ & \bar{D}_1 \end{bmatrix} \cdot \begin{bmatrix} V_1 \\ \bar{V}_1 \end{bmatrix}.$$

Note that

$$\begin{bmatrix} \mathbf{u}+\mathbf{v}\imath & \mathbf{u}-\mathbf{v}\imath \end{bmatrix} \cdot \begin{bmatrix} a+b\imath & \\ & a-b\imath \end{bmatrix} \cdot \begin{bmatrix} \mathbf{u}^{\mathsf{T}}+\mathbf{v}^{\mathsf{T}}\imath \\ \mathbf{u}^{\mathsf{T}}-\mathbf{v}^{\mathsf{T}}\imath \end{bmatrix} = \begin{bmatrix} \mathbf{u} & \mathbf{v} \end{bmatrix} \cdot 2 \begin{bmatrix} a & -b \\ -b & -a \end{bmatrix} \cdot \begin{bmatrix} \mathbf{u}^{\mathsf{T}} \\ \mathbf{v}^{\mathsf{T}} \end{bmatrix}.$$

Denote the jth column of V_1^{T} as $\mathbf{u}_j + \mathbf{v}_j\imath$ and the jth entry of the first column of D_1 is $a_j + b_j\imath$, where $\imath = \sqrt{-1}$ and $\mathbf{u}_j, \mathbf{v}_j, a_j, b_j$ are all real. Then

$$\begin{bmatrix} V_1^{\mathsf{T}} & V_2^{\mathsf{T}} \end{bmatrix} \cdot \begin{bmatrix} D_1 & \\ & D_2 \end{bmatrix} \cdot \begin{bmatrix} V_1 \\ V_2 \end{bmatrix}$$

$$= \begin{bmatrix} \mathbf{u}_1 & \mathbf{v}_1 & \cdots & \mathbf{u}_{k_1} & \mathbf{v}_{k_1} \end{bmatrix} \cdot \begin{bmatrix} \Lambda_1 & \Lambda_2 & \cdots & \Lambda_{k_1} \\ \Lambda_2 & \cdots & \Lambda_{k_1} & O \\ \vdots & \ddots & \ddots & \vdots \\ \Lambda_{k_l} & O & \cdots & O \end{bmatrix} \cdot \begin{bmatrix} \mathbf{u}_1^{\mathsf{T}} \\ \mathbf{v}_1^{\mathsf{T}} \\ \vdots \\ \mathbf{u}_{k_1}^{\mathsf{T}} \\ \mathbf{v}_{k_1}^{\mathsf{T}} \end{bmatrix},$$

where the 2-by-2 block Λ_j is

$$\Lambda_j = 2 \begin{bmatrix} a_j & -b_j \\ -b_j & -a_j \end{bmatrix}, \quad j = 1, 2, \ldots, k_1.$$

We perform the same transformations to all the conjugate eigenvalue pairs, then we obtain a real decomposition of the real Hankel matrix $H = \widehat{V}^{\mathsf{T}} \widehat{D} \widehat{V}$. Here, each diagonal block of $\widehat{D}$ corresponding to a real eigenvalue of C is an upper anti-triangular Hankel matrix, and each corresponding to a pair of conjugate eigenvalues is an upper anti-triangular block Hankel matrix with 2-by-2 blocks.

We claim that if the Hankel matrix H is positive semidefinite, then all the eigenvalues of C are real and of multiplicity one. This can be seen by recognizing that the following three cases of the diagonal blocks of $\widehat{D}$ cannot be positive semidefinite:

(1) an upper anti-triangular Hankel block whose size is larger than 1,

(2) a 2-by-2 block $\Lambda_j = 2 \begin{bmatrix} a_j & -b_j \\ -b_j & -a_j \end{bmatrix}$, and

(3) a block upper anti-triangular Hankel block with the blocks in case (2).

Therefore when a real rank-r Hankel matrix H is positive semidefinite and its leading $r \times r$ principal submatrix is positive definite, the block diagonal matrix $\widehat{D}$ in the generalized real Vandermonde decomposition must be diagonal. Hence this Hankel matrix admits a Vandermonde decomposition with r terms and all positive coefficients:

$$H = \sum_{k=1}^{r} \alpha_k \mathbf{v}_k \mathbf{v}_k^{\mathsf{T}}, \quad \alpha_k > 0, \quad \mathbf{v}_k = \begin{bmatrix} 1, \xi_k, \ldots, \xi_k^{n-1} \end{bmatrix}^{\mathsf{T}}.$$

This result for positive definite Hankel matrices is known [119, Lemma 0.2.1].

4.3.3 An Augmented Vandermonde Decomposition of Hankel Tensors

However, the associated Hankel matrix of a Hankel tensor is not necessarily with a nonsingular leading principal submatrix. Thus we shall study whether a positive semidefinite Hankel matrix can always decomposed into the form

$$H = \sum_{k=1}^{r-1} \alpha_k \mathbf{v}_k \mathbf{v}_k^\top + \alpha_r \mathbf{e}_n \mathbf{e}_n^\top, \quad \alpha_k \geq 0.$$

We first need a lemma about the rank-one modifications on a positive semidefinite matrix. Denote the range and the kernel of a matrix A as $\mathrm{Ran}(A)$ and $\mathrm{Ker}(A)$, respectively.

Lemma 4.4. *Let A be a positive semidefinite matrix with* $\mathrm{rank}(A) = r$. *Then there exists a unique $\alpha > 0$ such that $A - \alpha \mathbf{u}\mathbf{u}^\top$ is positive semidefinite with* $\mathrm{rank}(A - \alpha \mathbf{u}\mathbf{u}^\top) = r - 1$, *if and only if $\mathbf{u}$ is in the range of A.*

Proof. The condition $\mathrm{rank}(A - \alpha \mathbf{u}\mathbf{u}^\top) = \mathrm{rank}(A) - 1$ obviously indicates that $\mathbf{u} \in \mathrm{Ran}(A)$. Thus we only need to prove the "if" part of the statement.

Let the nonzero eigenvalues of A be $\lambda_1, \lambda_2, \ldots, \lambda_r$ and the corresponding eigenvectors be $\mathbf{x}_1, \mathbf{x}_2, \ldots, \mathbf{x}_r$, respectively. Since $\mathbf{u} \in \mathrm{Ran}(A)$, we can write $\mathbf{u} = \mu_1 \mathbf{x}_1 + \mu_2 \mathbf{x}_2 + \cdots + \mu_r \mathbf{x}_r$. Note that $\mathrm{rank}(A - \alpha \mathbf{u}\mathbf{u}^\top) = \mathrm{rank}(A) - 1$ also implies $\dim \mathrm{Ker}(A - \alpha \mathbf{u}\mathbf{u}^\top) = \dim \mathrm{Ker}(A) + 1$, and equivalently there exists a unique subspace $\mathrm{span}\{\mathbf{p}\}$ such that $A\mathbf{p} = \alpha \mathbf{u}(\mathbf{u}^\top \mathbf{p}) \neq \mathbf{0}$. Write $\mathbf{p} = \eta_1 \mathbf{x}_1 + \eta_2 \mathbf{x}_2 + \cdots + \eta_r \mathbf{x}_r$. Then there exists a unique linear combination and a unique scalar α satisfying

$$\eta_i = \mu_i / \lambda_i \ (i = 1, 2, \ldots, r), \quad \alpha = \left(\mu_1^2 / \lambda_1 + \cdots + \mu_r^2 / \lambda_r \right)^{-1}.$$

Then we verify the positive semidefiniteness of $A - \alpha \mathbf{u}\mathbf{u}^\top$. For an arbitrary $\mathbf{x} = \xi_1 \mathbf{x}_1 + \xi_2 \mathbf{x}_2 + \cdots + \xi_r \mathbf{x}_r$ in the range of A, we have

$$\mathbf{x}^\top A \mathbf{x} = \xi_1^2 \lambda_1 + \cdots + \xi_r^2 \lambda_r, \quad \mathbf{u}^\top \mathbf{x} = \mu_1 \xi_1 + \cdots + \mu_r \xi_r.$$

Along with the expression of α, the Hölder inequality indicates that $\mathbf{x}^\top A \mathbf{x} \geq \alpha (\mathbf{u}^\top \mathbf{x})^2$, that is, the rank-$(r-1)$ matrix $A - \alpha \mathbf{u}\mathbf{u}^\top$ is also positive semidefinite. $\square$

The following theorem tells that the leading $(r-1) \times (r-1)$ principal submatrix of a rank-r positive semidefinite Hankel matrix is always invertible, even when the leading $r \times r$ principal submatrix is rank deficient.

Theorem 4.5. *Let $H \in \mathbb{R}^{n \times n}$ be a positive semidefinite Hankel matrix with* $\mathrm{rank}(H) = r$. *If $H(:, n)$ is linearly dependent of the columns in $H(:, 1 : n-1)$, then the leading $r \times r$ principal submatrix $H(1 : r, 1 : r)$ is positive definite. If $H(:, n)$ is linearly independent of the columns in $H(:, 1 : n-1)$, then the leading $(r-1) \times (r-1)$ principal submatrix $H(1 : r-1, 1 : r-1)$ is positive definite.*

Proof. We apply the mathematical induction on the size n. First, the statement is apparently true for 2×2 positive semidefinite Hankel matrices. Assume that the statement holds for $(n-1) \times (n-1)$ Hankel matrices, then we consider the $n \times n$ case.

Case 1: When the last column $H(:,n)$ is linearly dependent of the columns in $H(:,1:n-1)$, the submatrix $H(1:n-1,1:n-1)$ is also a rank-r positive semidefinite Hankel matrix. Then from the induction hypothesis, $H(1:r,1:r)$ is full rank if $H(1:n-1,n-1)$ is linearly dependent of the columns in $H(1:n-1,1:n-2)$, and $H(1:r-1,1:r-1)$ is full rank otherwise. We shall show that the column $H(1:n-1,n-1)$ is always linearly dependent of the columns in $H(1:n-1,1:n-2)$.

Assuming on the contrary, the leading $(r-1) \times (r-1)$ principal submatrix $H(1:r-1,1:r-1)$ is positive definite, and the rank of $H(1:n-2,1:n-1)$ is $r-1$. Since the column $H(:,n)$ is linear dependent of the previous $(n-1)$ columns, the rank of $H(1:n-2,:)$ is also $r-1$. Thus the rectangular Hankel matrix $H(1:n-2,:)$ has a Vandermonde decomposition

$$H(1:n-2,:) = \sum_{k=1}^{r-1} \alpha_k \begin{bmatrix} 1 \\ \xi_k \\ \vdots \\ \xi_k^{n-3} \end{bmatrix} \begin{bmatrix} 1 & \xi_k & \cdots & \xi_k^{n-2} & \xi_k^{n-1} \end{bmatrix}.$$

Since $H(n-1,n-1) = H(n-2,n)$, the square Hankel matrix $H(1:n-1,1:n-1)$ has a corresponding decomposition

$$H(1:n-1,1:n-1) = \sum_{k=1}^{r-1} \alpha_k \begin{bmatrix} 1 \\ \xi_k \\ \vdots \\ \xi_k^{n-2} \end{bmatrix} \begin{bmatrix} 1 & \xi_k & \cdots & \xi_k^{n-2} \end{bmatrix}.$$

This contradicts with $\text{rank}(H(1:n-1,1:n-1)) = r$. Therefore $H(1:n-1,n-1)$ must be linearly dependent of the columns in $H(1:n-1,1:n-2)$. Hence the leading principal submatrix $H(1:r,1:r)$ is positive definite.

Case 2: When the last column $H(:,n)$ is linearly independent of the columns in $H(:,1:n-1)$, it is equivalent to that $\mathbf{e}_n$ is in the range of H. Thus from Lemma 4.4, there exists a scalar α_r such that $H - \alpha_r \mathbf{e}_n \mathbf{e}_n^\top$ is rank-$(r-1)$ and also positive semidefinite. Referring back to Case 1, we know that the leading principal submatrix $H(1:r-1,1:r-1)$ is positive definite. □

Following the above theorem, when $H(:,n)$ is linearly dependent of the columns in $H(:,1:n-1)$, the leading $r \times r$ principal submatrix $H(1:r,1:r)$ is positive definite. Thus H has a Vandermonde decomposition with positive coefficients:

$$H = \sum_{k=1}^{r} \alpha_k \mathbf{v}_k \mathbf{v}_k^\top, \quad \alpha_k > 0, \quad \mathbf{v}_k = \begin{bmatrix} 1, \xi_k, \ldots, \xi_k^{n-1} \end{bmatrix}^\top.$$

When $H(:,n)$ is linearly independent of the columns in $H(:,1:n-1)$, the leading $(r-1) \times (r-1)$ principal submatrix $H(1:r-1,1:r-1)$ is positive definite. Thus H has an augmented Vandermonde decomposition with positive coefficients:

$$H = \sum_{k=1}^{r-1} \alpha_k \mathbf{v}_k \mathbf{v}_k^\top + \alpha_r \mathbf{e}_n \mathbf{e}_n^\top, \quad \alpha_k > 0, \quad \mathbf{v}_k = \begin{bmatrix} 1, \xi_k, \ldots, \xi_k^{n-1} \end{bmatrix}^\top.$$

Combining the definition of strong Hankel tensors and the analysis at the end of Section 4.3.1, we arrive at the following theorem.

Theorem 4.6. *Let $\mathcal{H}$ be an mth-order n-dimensional Hankel tensor and the rank of its associated Hankel matrix be r. Then it is a strong Hankel tensor if and only if it admits a Vandermonde decomposition with positive coefficients:*

$$(4.6) \qquad \mathcal{H} = \sum_{k=1}^{r} \alpha_k \mathbf{v}_k^{\circ m}, \quad \alpha_k > 0, \quad \mathbf{v}_k = \left[1, \xi_k, \ldots, \xi_k^{n-1}\right]^{\top},$$

or an augmented Vandermonde decomposition with positive coefficients:

$$(4.7) \qquad \mathcal{H} = \sum_{k=1}^{r-1} \alpha_k \mathbf{v}_k^{\circ m} + \alpha_r \mathbf{e}_n^{\circ m}, \quad \alpha_k > 0, \quad \mathbf{v}_k = \left[1, \xi_k, \ldots, \xi_k^{n-1}\right]^{\top}.$$

After Theorem 4.6, the strong Hankel tensor cone is understood thoroughly. The polynomials induced by strong Hankel tensors are not only positive semidefinite and sum-of-squares, as proved in [102, Theorem 3.1] and [72, Corollary 2], but also sum-of-m^{th}-powers. The detailed algorithm for computing an augmented Vandermonde decomposition of a strong Hankel tensor is displayed as follows.

Example 4.2 ([112]). *An mth-order n-dimensional Hilbert tensor is defined by*

$$\mathcal{H}(i_1, i_2, \ldots, i_m) = \frac{1}{i_1 + i_2 + \cdots + i_m + 1}, \quad i_1, \ldots, i_m = 0, 1, \ldots, n-1.$$

Apparently, $\mathcal{H}$ is a special Hankel tensor with the generating vector $\left[1, \frac{1}{2}, \frac{1}{3}, \ldots, \frac{1}{mn-m+1}\right]^{\top}$. Moreover, its associated Hankel matrix is a Hilbert matrix, which is well known to be positive definite [112]. Thus a Hilbert tensor must be a strong Hankel tensor.

We take the 4th-order 5-dimensional Hilbert tensor, which is generated by $\left[1, \frac{1}{2}, \frac{1}{3}, \ldots, \frac{1}{17}\right]^{\top}$, as the second example. Applying Algorithm 4.2 and taking $\gamma = \frac{1}{18}$ in the algorithm, we obtain a standard Vandermonde decomposition of

$$\mathcal{H} = \sum_{k=1}^{9} \alpha_k \mathbf{v}_k^{\circ 4}, \quad \mathbf{v}_k = \left[1, \xi_k, \ldots, \xi_k^{n-1}\right]^{\top},$$

where α_k and ξ_k are displayed in Table 4.1.

k	1	2	3	4	5	6	7	8	9
ξ_k	0.9841	0.9180	0.8067	0.6621	0.5000	0.3379	0.1933	0.0820	0.0159
α_k	0.0406	0.0903	0.1303	0.1562	0.1651	0.1562	0.1303	0.0903	0.0406

Table 4.1: A Vandermonde decomposition for a Hilbert tensor.

From the computational result, we can see that a Hilbert tensor is actually a nonnegative strong Hankel tensor with a nonnegative Vandermonde decomposition.

Then we test some strong Hankel tensors without standard Vandermonde decompositions.

Algorithm 4.2 Augmented Vandermonde decomposition of a strong Hankel tensor.

Input: The generating vector $\mathbf{h}$ of a strong Hankel tensor

Output: Coefficients α_k; Poles ξ_k

1: Compute the Takagi factorization of the Hankel matrix H generated by $\mathbf{h}$:
$$H = UDU^{\top}$$

2: Recognize the rank r of H and whether $\mathbf{e}_n$ is in the range of U

3: **If** $r < n$, **then**

4:      **If** $\mathbf{e}_n \in \mathrm{Ran}(U)$, **then**

5:        $\xi_r = \mathrm{Inf}$

6:        $\alpha_r = \left(\sum_{j=1}^{r} U(n,j)^2 / D(j,j) \right)^{-1}$

7:        Apply Algorithm 4.2 for the strong Hankel tensor generated by $[h_0, \ldots, h_{mn-m-1}, h_{mn-m} - \alpha_r]$ to compute α_k and ξ_k for $k = 1, 2, \ldots, r-1$

8:      **ElseIf** $\mathbf{e}_n \notin \mathrm{Ran}(U)$, **then**

9:        $\mathbf{a} = U(1:r,1:r)^{-\top} D(1:r,1:r)^{-1} U(1:r,1:r)^{-1} \mathbf{h}(r:2r-1)$

10:      **EndIf**

11: **Else**

12:      $\mathbf{a} = U(1:r,1:r)^{-\top} D(1:r,1:r)^{-1} U(1:r,1:r)^{-1} [\mathbf{h}(r:2r-2)^{\top}, \gamma]^{\top}$, where γ is arbitrary

13: **EndIf**

14: Compute the roots $\xi_1, \xi_2, \ldots, \xi_r$ of the polynomial
$$p(\xi) = \xi^r - a_{r-1}\xi^{r-1} - \cdots - a_0\xi^0$$

15: Solve the Vandermonde system
$$\begin{bmatrix} 1 & 1 & \cdots & 1 \\ \xi_1 & \xi_2 & \cdots & \xi_r \\ \vdots & \vdots & \cdots & \vdots \\ \xi_1^{r-1} & \xi_2^{r-1} & \cdots & \xi_r^{r-1} \end{bmatrix} \begin{bmatrix} \alpha_1 \\ \alpha_2 \\ \vdots \\ \alpha_r \end{bmatrix} = \begin{bmatrix} h_0 \\ h_1 \\ \vdots \\ h_{r-1} \end{bmatrix}$$

16: **return** α_k, ξ_k for $k = 1, 2, \ldots, r$

Example 4.3. *Randomly generate two real scalars ξ_1, ξ_2. Then we construct a 4th-order n-dimensional strong Hankel tensor by*

$$
\mathcal{H} = \begin{bmatrix} 1 \\ 0 \\ \vdots \\ 0 \\ 0 \end{bmatrix}^{\circ 4} + \begin{bmatrix} 1 \\ \xi_1 \\ \vdots \\ \xi_1^{n-2} \\ \xi_1^{n-1} \end{bmatrix}^{\circ 4} + \begin{bmatrix} 1 \\ \xi_2 \\ \vdots \\ \xi_2^{n-2} \\ \xi_2^{n-1} \end{bmatrix}^{\circ 4} + \begin{bmatrix} 0 \\ 0 \\ \vdots \\ 0 \\ 1 \end{bmatrix}^{\circ 4}.
$$

For instance, we set the size $n = 10$ and apply Algorithm 4.2 to obtain an augmented Vandermonde decomposition of $\mathcal{H}$. We repeat this experiment $10{,}000$ times, and the average relative error between the computational solutions $\hat{\xi}_1, \hat{\xi}_2$ and the exact solutions ξ_1, ξ_2 is 4.7895×10^{-12}. That is, our algorithm recovers the augmented Vandermonde decomposition of $\mathcal{H}$ accurately.

4.3.4 The Second Inheritance Property of Hankel Tensors

Employing the augmented Vandermonde decomposition, we can prove the following theorem.

Theorem 4.7. *If a Hankel matrix has no negative (nonpositive, positive, or nonnegative) eigenvalues, then all of its associated higher-order Hankel tensors have no negative (nonpositive, positive, or nonnegative, respectively) H-eigenvalues.*

Proof. This statement for even order case is a direct corollary of Theorem 4.1.

Suppose that the Hankel matrix has no negative eigenvalues. When the order m is odd, decompose an mth-order strong Hankel tensor $\mathcal{H}$ into $\mathcal{H} = \sum_{k=1}^{r-1} \alpha_k \mathbf{v}_k^{\circ m} + \alpha_r \mathbf{e}_n^{\circ m}$ with $\alpha_k \geq 0$ $(k = 1, 2, \ldots, r)$. Then for an arbitrary vector $\mathbf{x}$, the first entry of $\mathcal{H}\mathbf{x}^{m-1}$ is

$$
(\mathcal{H}\mathbf{x}^{m-1})_1 = \sum_{k=1}^{r-1} (\mathbf{v}_k)_1 \cdot \alpha_k (\mathbf{v}_k^\top \mathbf{x})^{m-1} = \sum_{k=1}^{r-1} \alpha_k (\mathbf{v}_k^\top \mathbf{x})^{m-1} \geq 0.
$$

If $\mathcal{H}$ has no H-eigenvalues, then the statement is proved. Assume it has at least one H-eigenvalue λ, and let $\mathbf{x}$ be a corresponding H-eigenvector. Then when $x_1 \neq 0$, from the definition we have

$$
\lambda = (\mathcal{H}\mathbf{x}^{m-1})_1 / x_1^{m-1} \geq 0,
$$

since m is odd. When $x_1 = 0$, we know that $(\mathcal{H}\mathbf{x}^{m-1})_1$ must also be zero, thus all the item $\alpha_k (\mathbf{v}_k^\top \mathbf{x})^{m-1} = 0$ for $k = 1, 2, \ldots, r-1$. Therefore the tensor-vector product

$$
\mathcal{H}\mathbf{x}^{m-1} = \sum_{k=1}^{r-1} \mathbf{v}_k \cdot \alpha_k (\mathbf{v}_k^\top \mathbf{x})^{m-1} + \mathbf{e}_n \cdot \alpha_r x_n^{m-1} = \mathbf{e}_n \cdot \alpha_r x_n^{m-1},
$$

and it is apparent that $\mathbf{x} = \mathbf{e}_n$ and $\lambda = \alpha_r x_n^{m-1}$, where the H-eigenvalue λ is nonnegative. The other cases can be proved similarly. □

When the Hankel matrix has no negative eigenvalue, it is positive semidefinite, that is, the associated Hankel tensors are strong Hankel tensors, which may be of either even or odd order. Thus we have the following corollary.

Corollary 4.8 ([72]). *Strong Hankel tensors have no negative H-eigenvalues.*

We also have a quantified version of the second inheritance property.

Theorem 4.9. *Let $\mathcal{H}$ be an mth-order n-dimensional Hankel tensor, and H be its associated Hankel matrix. If H is positive semidefinite, then*

$$\lambda_{\min}(\mathcal{H}) \geq c \cdot \lambda_{\min}(H),$$

where $c = \min_{\mathbf{y} \in \mathbb{R}^n} \|\mathbf{y}^{*\frac{m}{2}}\|_2^2 / \|\mathbf{y}\|_m^m$ if m is even, and $c = \min_{\mathbf{y} \in \mathbb{R}^n} \|\mathbf{y}^{*\frac{m-1}{2}}\|_2^2 / \|\mathbf{y}\|_{m-1}^{m-1}$ if m is odd, both dependent on m and n. If H is negative semidefinite, then

$$\lambda_{\max}(\mathcal{H}) \leq c \cdot \lambda_{\max}(H).$$

Proof. When the minimal eigenvalue of H equals 0, the above equality holds for any nonnegative c. Moreover, when the order m is even, Theorem 4.2 produces the constant c. Thus we need only to discuss the situation that H is positive definite and m is odd.

Since H is positive definite, the Hankel tensor $\mathcal{H}$ has a standard Vandermonde decomposition with positive coefficients:

$$\mathcal{H} = \sum_{k=1}^{r} \alpha_k \mathbf{v}_k^{\circ m}, \quad \alpha_k > 0, \quad \mathbf{v}_k = \left[1, \xi_k, \ldots, \xi_k^{n-1}\right]^{\top},$$

where ξ_k are mutually distinct. Then by the proof of Theorem 4.7, for an arbitrary nonzero $\mathbf{x} \in \mathbb{R}^n$,

$$(\mathcal{H}\mathbf{x}^{m-1})_1 = \sum_{k=1}^{r} \alpha_k (\mathbf{v}_k^{\top}\mathbf{x})^{m-1} > 0,$$

since $[\mathbf{v}_1, \mathbf{v}_2, \ldots, \mathbf{v}_r]$ spans the whole space. If λ and $\mathbf{x}$ are an H-eigenvalue and a corresponding H-eigenvector of $\mathcal{H}$, then λ must be positive and the first entry x_1 of $\mathbf{x}$ must be nonzero.

Let $\mathbf{h} \in \mathbb{R}^{m(n-1)+1}$ be the generating vector of both the Hankel matrix H and the Hankel tensor $\mathcal{H}$. Denote $\mathbf{h}_1 = \mathbf{h}(1 : (m-1)(n-1)+1)$, which generates a Hankel matrix H_1 and an $(m-1)^{\text{st}}$-order Hankel tensor $\mathcal{H}_1$. Note that H_1 is a leading principal submatrix of H and $\mathcal{H}_1$ is exactly the first row tensor of $\mathcal{H}$, that is, $\mathcal{H}(1, :, :, \ldots, :)$. Then we have

$$\lambda = \frac{(\mathcal{H}\mathbf{x}^{m-1})_1}{x_1^{m-1}} = \frac{\mathcal{H}_1\mathbf{x}^{m-1}}{x_1^{m-1}} \geq \frac{\mathcal{H}_1\mathbf{x}^{m-1}}{\|\mathbf{x}\|_{m-1}^{m-1}}.$$

Now $m - 1$ is an even number, so we know from Theorem 4.2 that there is a constant c such that $\lambda_{\min}(\mathcal{H}_1) \geq c \cdot \lambda_{\min}(H_1)$. Therefore for each existing H-eigenvalue λ of $\mathcal{H}$, we obtain

$$\lambda \geq \lambda_{\min}(\mathcal{H}_1) \geq c \cdot \lambda_{\min}(H_1) \geq c \cdot \lambda_{\min}(H).$$

The last inequality holds because H_1 is a principal submatrix of H [49, Theorem 8.1.7]. □

It is unclear whether we have a similar quantified form of the extremal H-eigenvalues of a Hankel tensor when its associated Hankel matrix has both positive and negative eigenvalues.

4.4 The Third Inheritance Property of Hankel Tensors

We have proved two inheritance properties of Hankel tensors in this chapter. We now raise the third inheritance property of Hankel tensors as conjecture.

Conjecture If a lower-order Hankel tensor has no negative H-eigenvalues, then its associated higher-order Hankel tensor with the same generating vector, where the higher order is a multiple of the lower order, inherits the same property.

We see that the first inheritance property of Hankel tensors established in this chapter is only a special case of this inheritance property, that is, the lower-order Hankel tensor is of even order. At this moment, we are unable to prove or disprove this conjecture if the lower-order Hankel tensor is of odd order. However, if this conjecture is true, then it is of significance. If the lower-order Hankel tensor is of odd order while the higher-order Hankel tensor is of even order, then the third inheritance property would provide a new way to identify some positive semidefinite Hankel tensors, as well as a link between odd-order symmetric tensors without negative H-eigenvalues and positive semidefinite symmetric tensors.

Part III

$\mathcal{M}$-Tensors

Chapter 5

Definitions and Basic Properties

The sign pattern of matrix or tensor components is an interesting property. Both nonnegative and $\mathcal{M}$-tensors in Chapters 5 and 6 have special sign patterns and spectral properties. This chapter is devoted to basic discussions of $\mathcal{M}$-tensors, and some further results related to multilinear systems of equations will be presented in the next chapter.

M-matrix is an important class of matrices and has been well studied. It is closely related to spectral graph theory, the numerical solutions of PDEs, the stationary distribution of Markov chains, and the convergence of iterative methods for linear systems. There are no less than 50 equivalent definitions of nonsingular M-matrices (eg, [9, Chapter 6]). How can they be generalized into the higher-order case? Are they still the equivalent conditions of higher-order nonsingular $\mathcal{M}$-tensors? What properties of $\mathcal{M}$-tensors can they indicate? We answer these questions in this and subsequent chapters.

Zhang et al. [133] first extended M-matrices to $\mathcal{M}$-tensors and studied their spectral properties. The main result in their paper is that every eigenvalue of a nonsingular $\mathcal{M}$-tensor has a positive real part, which is the same as the M-matrix. Ding et al. [38] attempted to extend other equivalent definitions of nonsingular M-matrices (semi-positivity, monotonicity, etc.) to the tensor case. He and Huang [51] studied some inequalities for the eigenvalues of $\mathcal{M}$-tensors. Furthermore, Ding and Wei [42] investigated the multilinear systems of equations whose coefficient tensors are nonsingular $\mathcal{M}$-tensors.

5.1 Preliminaries

We present some preliminaries associated with the Perron-Frobenius theorem for nonnegative tensors and several well-known results for M-matrices in this section. Then we introduce the definitions of $\mathcal{M}$-tensors.

5.1.1 Nonnegative Tensor

Because of the difficulties in studying the properties of a general tensor, researchers focus on selected structured tensors. The nonnegative tensor with

Theory and Computation of Tensors.
http://dx.doi.org/10.1016/B978-0-12-803953-3.50005-8

nonnegative components is one of the most well-studied tensors. The Perron-Frobenius theorem is the most famous result for nonnegative matrices (cf. [9]), which investigates their spectral radii and the corresponding nonnegative eigenvectors. Researchers also propose several similar results and refer to them as the Perron-Frobenius theorem for nonnegative tensors. The spectral radius of a nonnegative tensor $\mathcal{B}$ is denoted by

$$\rho(\mathcal{B}) = \max\left\{|\lambda| \,\middle|\, \lambda \text{ is an eigenvalue of } \mathcal{B}\right\}.$$

Before stating the Perron-Frobenius theorem, we introduce the irreducible and weakly irreducible tensors, both of which are generalizations of irreducible matrices.

Definition 5.1 (Irreducible Tensor [19]). *A tensor $\mathcal{B}$ is called* reducible *if there exists a nonempty proper index subset $I \subset \{1, 2, \dots, n\}$ such that*

$$b_{i_1 i_2 \dots i_m} = 0, \ \forall i_1 \in I, \ \forall i_2, i_3, \dots, i_m \notin I.$$

Otherwise, we say $\mathcal{B}$ is irreducible.

Definition 5.2 (Weakly Irreducible Nonnegative Tensor [48]). *We call a nonnegative matrix $GM(\mathcal{B})$ the* representation *associated with a nonnegative tensor $\mathcal{B}$ with the (i, j)-th entry of $GM(\mathcal{B})$ defined to be the summation of $b_{i i_2 i_3 \dots i_m}$ with indices $\{i_2, i_3, \dots, i_m\} \ni j$. We call a tensor $\mathcal{B}$* weakly reducible, *if its representation $GM(\mathcal{B})$ is reducible. If $\mathcal{B}$ is not weakly reducible, then it is called* weakly irreducible.

Now with these concepts, we can recall several results which we group under the Perron-Frobenius theorem for nonnegative tensors.

Theorem 5.1 (The Perron-Frobenius Theorem for Nonnegative Tensors). *If $\mathcal{B}$ is a nonnegative tensor of order m and dimension n, then $\rho(\mathcal{B})$ is an eigenvalue of $\mathcal{B}$ with a nonnegative eigenvector $x \in \mathbb{R}^n_+$ [129].*

If $\mathcal{B}$ is strictly nonnegative, then $\rho(\mathcal{B}) > 0$ [56].

If in addition $\mathcal{B}$ is weakly irreducible, then $\rho(\mathcal{B})$ is an eigenvalue of $\mathcal{B}$ with a positive eigenvector $x \in \mathbb{R}^n_{++}$ [48].

Suppose that furthermore $\mathcal{B}$ is irreducible, and λ is an eigenvalue with a nonnegative eigenvector, then $\lambda = \rho(\mathcal{B})$ [19].

5.1.2   From M-Matrix to $\mathcal{M}$-Tensor

We briefly introduce some essential properties of M-matrices in this section as a benchmark of our discussion of $\mathcal{M}$-tensors.

A matrix is called a *Z-matrix* if all its off-diagonal entries are nonpositive. It is apparent that a Z-matrix A can be written as

$$A = sI - B,$$

where B is a nonnegative matrix; when $s \geq \rho(B)$, we call A as an *M-matrix*; furthermore, when $s > \rho(B)$, we call A as a *nonsingular M-matrix*.

There are more than 50 conditions in the literature that are equivalent to the definition of nonsingular M-matrix. We list just 12 of them here, which will be involved in these two chapters (see [9, Chapter 6] for details).

Theorem 5.2. *If A is a Z-matrix, then the following conditions are equivalent:*

(C1) *A is a nonsingular M-matrix;*

(C2) *A + D is nonsingular for any nonnegative diagonal matrix D;*

(C3) *Every real eigenvalue of A is positive;*

(C4) *A is* positive stable; *that is, the real part of each eigenvalue of A is positive;*

(C5) *A is* semi-positive; *that is, there exists* $\mathbf{x} > \mathbf{0}$ *with* $A\mathbf{x} > \mathbf{0}$;

(C6) *There exists* $\mathbf{x} \geq \mathbf{0}$ *with* $A\mathbf{x} > \mathbf{0}$;

(C7) *A has all positive diagonal entries and there exists a positive diagonal matrix D such that AD is strictly diagonally dominant;*

(C8) *A has all positive diagonal entries and there exists a positive diagonal matrix D such that DAD is strictly diagonally dominant;*

(C9) *A is* monotone; *that is,* $A\mathbf{x} \geq \mathbf{0}$ *implies* $\mathbf{x} \geq \mathbf{0}$;

(C10) *There exists an inverse-positive matrix B and a nonsingular M-matrix C such that A = BC;*

(C11) *A has a* convergent regular splitting; *that is,* $A = M - N$, *where* $M^{-1} \geq 0$, $N \geq 0$, *and* $\rho(M^{-1}N) < 1$;

(C12) *A does not reverse the sign of any vector; that is, if* $\mathbf{x} \neq \mathbf{0}$ *and* $\mathbf{b} = A\mathbf{x}$, *then for some subscript i,* $x_i \cdot b_i > 0$.

In order to present the concept of $\mathcal{M}$-tensors, Zhang et al. [133] first define the $\mathcal{Z}$-tensor, similar to the matrix case. Recall that the unit tensor $\mathcal{I}$ is the tensor with all diagonal entries being 1 and off-diagonal entries being 0.

Definition 5.3 ($\mathcal{Z}$-tensor). *We call $\mathcal{A}$ a $\mathcal{Z}$-tensor, if all of its off-diagonal entries are nonpositive, which is equivalent to $\mathcal{A} = s\mathcal{I} - \mathcal{B}$, where $\mathcal{B}$ is a nonnegative tensor ($\mathcal{B} \geq 0$).*

Definition 5.4 ($\mathcal{M}$-tensor). *We call a $\mathcal{Z}$-tensor $\mathcal{A} = s\mathcal{I} - \mathcal{B}$ ($\mathcal{B} \geq 0$) as an $\mathcal{M}$-tensor if $s \geq \rho(\mathcal{B})$; we call it as a* nonsingular $\mathcal{M}$-tensor *(or strong $\mathcal{M}$-tensor in some literature) if $s > \rho(\mathcal{B})$.*

5.2   Spectral Properties of $\mathcal{M}$-Tensors

An essential property of M-matrices is that they are positive stable, that is, all the eigenvalues are on the right half plane. The positivity of eigenvalues implies nonsingularity, stability, and positive definiteness. The following theorem tells us that $\mathcal{M}$-tensors have the same spectral composition.

Theorem 5.3. *Let $\mathcal{A}$ be a $\mathcal{Z}$-tensor. Then $\mathcal{A}$ is a nonsingular $\mathcal{M}$-tensor if and only if the real part of each eigenvalue of $\mathcal{A}$ is positive.*

Proof. If $\mathcal{A}$ is a nonsingular $\mathcal{M}$-tensor, then $\mathcal{A} = s\mathcal{I} - \mathcal{B}$, where $\mathcal{B}$ is nonnegative and $s > \rho(\mathcal{B})$. Denote λ as an eigenvalue of $\mathcal{A}$. Then $s - \lambda$ is an eigenvalue of $\mathcal{B}$, thus $|s - \lambda| \leq \rho(\mathcal{B}) < s$. Therefore the real part

$$\mathrm{Re}\lambda = s - \mathrm{Re}(s - \lambda) \geq s - |s - \lambda| > 0.$$

If the real part of each eigenvalue of $\mathcal{A} = s\mathcal{I} - \mathcal{B}$ is positive, then $s > \mathrm{Re}\mu$, where μ is an arbitrary eigenvalue of the nonnegative tensor $\mathcal{B}$. By the Perron-Frobenious theorem for nonnegative tensors, the spectral radius $\rho(\mathcal{B})$ is exactly a real eigenvalue of $\mathcal{B}$. Take $\mu = \rho(\mathcal{B})$, and we get $s > \rho(\mathcal{B})$. Hence $\mathcal{A}$ is a nonsingular $\mathcal{M}$-tensor. $\square$

Furthermore, the Perron-Frobenious theorem for nonnegative tensors says that the spectral radius of a nonnegative tensor is not only a real eigenvalue but also an H-eigenvalue, that is, an eigenvalue with a real eigenvector, and an H^+-eigenvalue, that is, an eigenvalue with a nonnegative eigenvector. Thereby, we obtain another three equivalent definitions of nonsingular $\mathcal{M}$-tensors.

Corollary 5.4. *If $\mathcal{A}$ is a $\mathcal{Z}$-tensor, then the following conditions are equivalent:*

- *$\mathcal{A}$ is a nonsingular $\mathcal{M}$-tensor;*

- *Every real eigenvalue of $\mathcal{A}$ is positive;*

- *Every H-eigenvalue of $\mathcal{A}$ is positive;*

- *Every H^+-eigenvalue of $\mathcal{A}$ is positive.*

Recall that an even order symmetric tensor $\mathcal{A}$ is called *positive definite* if $\mathcal{A}x^m > 0$ for an arbitrary nonzero $\mathbf{x} \in \mathbb{R}^n$. Qi [98] proved that a tensor is positive definite if and only if all of its H-eigenvalues are positive. Thus we obtain the following equivalent definition for even-order symmetric $\mathcal{M}$-tensors.

Corollary 5.5. *Let $\mathcal{A}$ be an even-order symmetric $\mathcal{Z}$-tensor. Then $\mathcal{A}$ is a nonsingular $\mathcal{M}$-tensor if and only if $\mathcal{A}$ is positive definite.*

5.3   Semi-Positivity

In this section, we will propose an equivalent definition of nonsingular $\mathcal{M}$-tensors following conditions (C5) and (C6) in Section 5.1.2.

5.3.1   Definitions

First, we extend *semi-positivity* from matrices [9] to tensors.

Definition 5.5 (Semi-positive tensor). *We call $\mathcal{A}$ a semi-positive tensor, if there exists $\mathbf{x} > \mathbf{0}$ such that $\mathcal{A}x^{m-1} > \mathbf{0}$.*

Because of the continuity of the tensor-vector product on the entries of the vector, the requirement $\mathbf{x} > \mathbf{0}$ in the definition can be relaxed into $\mathbf{x} \geq \mathbf{0}$. We verify this statement as follows.

Theorem 5.6. *A tensor $\mathcal{A}$ is semi-positive if and only if there exists $\mathbf{x} \geq \mathbf{0}$ such that $\mathcal{A}x^{m-1} > \mathbf{0}$.*

Proof. Define a map $T_{\mathcal{A}}(\mathbf{x}) = (\mathcal{A}\mathbf{x}^{m-1})^{[\frac{1}{m-1}]}$, then $\mathbf{x} \mapsto T_{\mathcal{A}}(\mathbf{x})$ is continuous and bounded [21].

If $\mathcal{A}$ is semi-positive, then it is trivial that there is $\mathbf{x} \geq \mathbf{0}$ such that $\mathcal{A}\mathbf{x}^{m-1} > \mathbf{0}$, according to the definition.

If there exists $\mathbf{x} \geq \mathbf{0}$ with $\mathcal{A}\mathbf{x}^{m-1} > \mathbf{0}$, then there must be a closed ball $B(T_{\mathcal{A}}(\mathbf{x}), \varepsilon)$ in $\mathbb{R}^n_{++}$, where $B(\mathbf{c}, r) := \{\mathbf{v} \mid \|\mathbf{v} - \mathbf{c}\| \leq r\}$. Since $T_{\mathcal{A}}$ is continuous, there exists $\delta > 0$ such that $T_{\mathcal{A}}(\mathbf{y}) \in B(T_{\mathcal{A}}(\mathbf{x}), \varepsilon)$ for all $\mathbf{y} \in B(\mathbf{x}, \delta)$. Let $\mathbf{d}$ be a zero-one vector with $d_i = 1$ if $x_i = 0$ and $d_i = 0$ if $x_i > 0$. Take $\mathbf{y} = \mathbf{x} + \frac{\delta}{\|\mathbf{d}\|}\mathbf{d} \in B(\mathbf{x}, \delta)$, then $\mathbf{y} > \mathbf{0}$ and $T_{\mathcal{A}}(\mathbf{y}) > \mathbf{0}$. Therefore $\mathcal{A}$ is semi-positive. $\qquad\square$

It is well known that a Z-matrix is a nonsingular M-matrix if and only if it is semi-positive, that is, the equivalent definition (C5). Furthermore, we will draw a similar conclusion for nonsingular $\mathcal{M}$-tensors. The proof of this theorem will be presented at the end of this section after studying some properties of semi-positive $\mathcal{Z}$-tensors.

Theorem 5.7. *A $\mathcal{Z}$-tensor is a nonsingular $\mathcal{M}$-tensor if and only if it is semi-positive.*

5.3.2 Semi-Positive $\mathcal{Z}$-Tensors

The first property concerns the diagonal entries of a semi-positive $\mathcal{Z}$-tensor.

Proposition 5.8. *A semi-positive $\mathcal{Z}$-tensor has positive diagonal entries.*

Proof. When $\mathcal{A}$ is a semi-positive $\mathcal{Z}$-tensor, there exists $\mathbf{x} > \mathbf{0}$ such that $\mathcal{A}\mathbf{x}^{m-1} > \mathbf{0}$. Consequently, we have

$$(\mathcal{A}\mathbf{x}^{m-1})_i = a_{ii\ldots i}x_i^{m-1} + \sum_{(i_2,i_3,\ldots,i_m) \neq (i,i,\ldots,i)} a_{ii_2\ldots i_m}x_{i_2}\ldots x_{i_m} > 0,$$

for $i = 1, 2, \ldots, n$. From $x_j > 0$ and $a_{ii_2 i_3 \ldots i_m} \leq 0$ for $(i_2, i_3, \ldots, i_m) \neq (i, i, \ldots, i)$, we conclude that $a_{ii\ldots i} > 0$ for $i = 1, 2, \ldots, n$. $\qquad\square$

We have a series of equivalent conditions for semi-positive $\mathcal{Z}$-tensors, following conditions (C7), (C8), (C10), and (C11) in Section 5.1.2.

Proposition 5.9. *A $\mathcal{Z}$-tensor $\mathcal{A}$ is semi-positive if and only if $\mathcal{A}$ has positive diagonal entries and there exists a positive diagonal matrix D such that $\mathcal{A}D^{m-1}$ is strictly diagonally dominant.*

Proof. Let $D = \text{diag}(d_1, d_2, \ldots, d_n)$. Then $\mathcal{A}D^{m-1}$ is strictly diagonally dominant means

$$\left| a_{ii\ldots i}d_i^{m-1} \right| > \sum_{(i_2,i_3,\ldots,i_m) \neq (i,i,\ldots,i)} \left| a_{ii_2\ldots i_m}d_{i_2}\ldots d_{i_m} \right|, \quad i = 1, 2, \ldots, n.$$

If $\mathcal{A}$ is a semi-positive $\mathcal{Z}$-tensor, then $a_{ii\ldots i} > 0$ for $i = 1, 2, \ldots, n$ from Proposition 5.8, $a_{ii_2\ldots i_m} \leq 0$ for $(i_2, i_3, \ldots, i_m) \neq (i, i, \ldots, i)$, and there is $\mathbf{x} > \mathbf{0}$ with $\mathcal{A}\mathbf{x}^{m-1} > \mathbf{0}$. Let $D = \text{diag}(\mathbf{x})$, and we conclude that D is positive diagonal and $\mathcal{A}D^{m-1}$ is strictly diagonally dominant.

If $\mathcal{A}$ has positive diagonal entries and there exists a positive diagonal matrix D such that AD^{m-1} is strictly diagonally dominant, let $\mathbf{x} = \mathrm{diag}(D) > \mathbf{0}$, then $\mathcal{A}\mathbf{x}^{m-1} > \mathbf{0}$ since $a_{ii...i} > 0$ for $i = 1, 2, \ldots, n$ and $a_{ii_2...i_m} \leq 0$ for $(i_2, i_3, \ldots, i_m) \neq (i, i, \ldots, i)$. Thus $\mathcal{A}$ is a semi-positive tensor. $\qquad\square$

Proposition 5.10. *A $\mathcal{Z}$-tensor $\mathcal{A}$ is semi-positive if and only if $\mathcal{A}$ has all positive diagonal entries and there exist two positive diagonal matrices D_1 and D_2 such that $D_1 A D_2^{m-1}$ is strictly diagonally dominant.*

Proof. Notice that $D_1 A D_2^{m-1}$ is strictly diagonally dominant if and only if $A D_2^{m-1}$ is strictly diagonally dominant because of the positivity of D_1. This proposition is a corollary of Proposition 5.9. $\qquad\square$

For a diagonal tensor $\mathcal{D}$ and another tensor $\mathcal{A}$ of the same size, we sometimes denote their composite as

$$(\mathcal{D}\mathcal{A})_{i_1 i_2 ... i_m} = d_{i_1 i_1 ... i_1} \cdot a_{i_1 i_2 ... i_m},$$

which indicates that $(\mathcal{D}\mathcal{A})\mathbf{x}^{m-1} = \mathcal{D}\left((\mathcal{A}\mathbf{x}^{m-1})^{\left[\frac{1}{m-1}\right]}\right)^{m-1}$. The inverse of the diagonal tensor $\mathcal{D}$ without zero diagonal entries is diagonal with

$$(\mathcal{D}^{-1})_{i_1 i_1 ... i_1} = d_{i_1 i_1 ... i_1}^{-1},$$

which implies $(\mathcal{D}^{-1}\mathcal{D})\mathbf{x}^{m-1} = (\mathcal{D}\mathcal{D}^{-1})\mathbf{x}^{m-1} = \mathcal{I}\mathbf{x}^{m-1} = \mathbf{x}^{[m-1]}$.

Proposition 5.11. *A $\mathcal{Z}$-tensor $\mathcal{A}$ is semi-positive if and only if there exist a positive diagonal tensor $\mathcal{D}$ and a semi-positive $\mathcal{Z}$-tensor $\mathcal{C}$ such that $\mathcal{A} = \mathcal{D}\mathcal{C}$.*

Proof. Let $\mathcal{D}$ be the diagonal part of $\mathcal{A}$, that is, it is diagonal with $d_{ii...i} = a_{ii...i}$ for all i, and $\mathcal{C} = \mathcal{D}^{-1}\mathcal{A}$. Clearly, $\mathcal{A} = \mathcal{D}\mathcal{C}$.

If $\mathcal{A}$ is a semi-positive $\mathcal{Z}$-tensor, then $\mathcal{D}$ is positive diagonal and there exists $\mathbf{x} > \mathbf{0}$ with $\mathcal{A}\mathbf{x}^{m-1} > \mathbf{0}$. Then the vector $\mathcal{C}\mathbf{x}^{m-1} = \mathcal{D}^{-1}\left((\mathcal{A}\mathbf{x}^{m-1})^{\left[\frac{1}{m-1}\right]}\right)^{m-1}$ is also positive. Therefore $\mathcal{C}$ is also a semi-positive $\mathcal{Z}$-tensor.

If $\mathcal{C}$ is a semi-positive $\mathcal{Z}$-tensor and $\mathcal{D}$ is positive diagonal, then there exists $\mathbf{x} > \mathbf{0}$ with $\mathcal{C}\mathbf{x}^{m-1} > \mathbf{0}$. Then the vector $\mathcal{A}\mathbf{x}^{m-1} = \mathcal{D}\left((\mathcal{C}\mathbf{x}^{m-1})^{\left[\frac{1}{m-1}\right]}\right)^{m-1}$ is also positive. Thus $\mathcal{A}$ is a semi-positive $\mathcal{Z}$-tensor. $\qquad\square$

Remark 5.12. *After we prove Theorem 5.7, the above proposition can be restated as: A $\mathcal{Z}$-tensor $\mathcal{A}$ is a nonsingular $\mathcal{M}$-tensor if and only if there exist a positive diagonal tensor $\mathcal{D}$ and a nonsingular $\mathcal{M}$-tensor $\mathcal{C}$ with $\mathcal{A} = \mathcal{D}\mathcal{C}$.*

Proposition 5.13. *A $\mathcal{Z}$-tensor $\mathcal{A}$ is semi-positive if and only if there exist a positive diagonal tensor $\mathcal{D}$, a nonnegative tensor $\mathcal{E}$ such that $\mathcal{A} = \mathcal{D} - \mathcal{E}$, and $\mathbf{x} > \mathbf{0}$ such that $(\mathcal{D}^{-1}\mathcal{E})\mathbf{x}^{m-1} < \mathbf{x}^{[m-1]}$.*

Proof. Let $\mathcal{D}$ be the diagonal part of $\mathcal{A}$ and $\mathcal{E} = \mathcal{D} - \mathcal{A}$. Clearly, $\mathcal{A} = \mathcal{D} - \mathcal{E}$ and $\mathcal{D}^{-1}\mathcal{E} = \mathcal{I} - \mathcal{D}^{-1}\mathcal{A}$.

If $\mathcal{A}$ is a semi-positive $\mathcal{Z}$-tensor, then $\mathcal{D}$ is positive and there exists $\mathbf{x} > \mathbf{0}$ with $\mathcal{A}\mathbf{x}^{m-1} > \mathbf{0}$. Then $\mathcal{D}\mathbf{x}^{m-1} > \mathcal{E}\mathbf{x}^{m-1}$, and $(\mathcal{D}^{-1}\mathcal{E})\mathbf{x}^{m-1} < \mathbf{x}^{[m-1]}$.

If there exists $\mathbf{x} > \mathbf{0}$ with $(\mathcal{D}^{-1}\mathcal{E})\mathbf{x}^{m-1} < \mathbf{x}^{[m-1]}$, then $\mathcal{E}\mathbf{x}^{m-1} < \mathcal{D}\mathbf{x}^{m-1}$, and $\mathcal{A}\mathbf{x}^{m-1} > \mathbf{0}$. Therefore $\mathcal{A}$ is a semi-positive $\mathcal{Z}$-tensor. $\qquad\square$

Remark 5.14. *It follows from [129, Lemma 5.4] that a semi-positive $\mathcal{Z}$-tensor can be split into $\mathcal{A} = \mathcal{D} - \mathcal{E}$, where $\mathcal{D}$ is a positive diagonal tensor and $\mathcal{E}$ is a nonnegative tensor with $\rho(\mathcal{D}^{-1}\mathcal{E}) < 1$.*

Next we present some examples of nontrivial semi-positive $\mathcal{Z}$-tensors.

Proposition 5.15. *A strictly diagonally dominant $\mathcal{Z}$-tensor with nonnegative diagonal entries is a semi-positive $\mathcal{Z}$-tensor.*

Proof. Use **e** to denote the vector of all ones. It is easy to show, for a $\mathcal{Z}$-tensor $\mathcal{A}$ with all nonnegative diagonal entries that strictly diagonal dominance is equivalent to $\mathcal{A}\mathbf{e}^{m-1} > \mathbf{0}$. Since $\mathbf{e} > \mathbf{0}$, the result follows from the definition of semi-positive $\mathcal{Z}$-tensors. □

Proposition 5.16. *A weakly irreducible nonsingular $\mathcal{M}$-tensor is a semi-positive $\mathcal{Z}$-tensor.*

Proof. When $\mathcal{A}$ is a nonsingular $\mathcal{M}$-tensor, we write $\mathcal{A} = s\mathcal{I} - \mathcal{B}$, where $\mathcal{B} \geq 0$ and $s > \rho(\mathcal{B})$. Since $\mathcal{A}$ is weakly irreducible, so is $\mathcal{B}$. Then there exists $\mathbf{x} > \mathbf{0}$ such that $\mathcal{B}\mathbf{x}^{m-1} = \rho(\mathcal{B})\mathbf{x}^{[m-1]}$ from the Perron-Frobenius Theorem for nonnegative tensors (cf. Theorem 5.1), and

$$\mathcal{A}\mathbf{x}^{m-1} = (s - \rho(\mathcal{B}))\mathbf{x}^{[m-1]} > 0.$$

Therefore $\mathcal{A}$ is a semi-positive tensor. □

5.3.3 Proof of Theorem 5.7

Our aim is to prove the equality

$$\{\text{semi-positive } \mathcal{Z}\text{-tensors}\} = \{\text{nonsingular } \mathcal{M}\text{-tensors}\}.$$

The first step is to verify the "⊆" part, which is relatively simple.

Lemma 5.17. *A semi-positive $\mathcal{Z}$-tensor is also a nonsingular $\mathcal{M}$-tensor.*

Proof. When $\mathcal{A}$ is semi-positive, there exists $\mathbf{x} > \mathbf{0}$ with $\mathcal{A}\mathbf{x}^{m-1} > \mathbf{0}$. We write $\mathcal{A} = s\mathcal{I} - \mathcal{B}$ since $\mathcal{A}$ is a $\mathcal{Z}$-tensor, where $\mathcal{B} \geq 0$. Then

$$\min_{1 \leq i \leq n} \frac{(\mathcal{B}\mathbf{x}^{m-1})_i}{x_i^{m-1}} \leq \rho(\mathcal{B}) \leq \max_{1 \leq i \leq n} \frac{(\mathcal{B}\mathbf{x}^{m-1})_i}{x_i^{m-1}}.$$

Thus $s - \rho(\mathcal{B}) \geq s - \max\limits_{1 \leq i \leq n} \dfrac{(\mathcal{B}\mathbf{x}^{m-1})_i}{x_i^{m-1}} = \min\limits_{1 \leq i \leq n} \dfrac{(\mathcal{A}\mathbf{x}^{m-1})_i}{x_i^{m-1}} > 0$, since both $\mathbf{x}$ and $\mathcal{A}\mathbf{x}^{m-1}$ are positive. Therefore $\mathcal{A}$ is a nonsingular $\mathcal{M}$-tensor. □

Next we prove the ⊇ part employing a partition of general nonnegative tensors into weakly irreducible leading subtensors.

Lemma 5.18. *A nonsingular $\mathcal{M}$-tensor is a semi-positive $\mathcal{Z}$-tensor.*

Proof. Assume that a nonsingular $\mathcal{M}$-tensor $\mathcal{A} = s\mathcal{I} - \mathcal{B}$ is weakly reducible, or we have proved a weakly irreducible nonsingular $\mathcal{M}$-tensor is also a semi-positive $\mathcal{Z}$-tensor in Proposition 5.16. Then $\mathcal{B}$ is also weakly reducible. Following [56, Theorem 5.2], the index set $I = \{1, 2, \ldots, n\}$ can be partitioned into $I = I_1 \sqcup I_2 \sqcup \cdots \sqcup I_k$ (here $A = A_1 \sqcup A_2$ means that $A = A_1 \cup A_2$ and $A_1 \cap A_2 = \varnothing$) such that

(1) $\mathcal{B}_{I_t I_t \ldots I_t}$ is weakly irreducible,

(2) $b_{i_1 i_2 \ldots i_m} = 0$ for $i_1 \in I_t$ and $\{i_2, i_3, \ldots, i_m\} \not\subseteq I_t \sqcup I_{t+1} \sqcup \cdots \sqcup I_k$, $t = 1, 2, \ldots, k$.

Without loss of generality, we can assume that

$$I_1 = \{1, 2, \ldots, n_1\},$$
$$I_2 = \{n_1 + 1, n_1 + 2, \ldots, n_2\},$$
$$\cdots \quad \cdots \quad \cdots$$
$$I_k = \{n_{k-1} + 1, n_{k-1} + 2, \ldots, n\}.$$

We introduce the following denotations

$$\mathcal{B}_{(t,\mathbf{a})} := \mathcal{B}_{I_t I_{a_1} \ldots I_{a_{m-1}}}$$

and

$$\mathcal{B}_{(t,\mathbf{a})} \mathbf{z}_{\mathbf{a}}^{m-1} := \mathcal{B}_{(t,\mathbf{a})} \times_{a_1} \mathbf{z}_{a_1} \times \cdots \times_{a_{m-1}} \mathbf{z}_{a_{m-1}},$$

where $\mathbf{a}$ is an index vector of length $m - 1$ and $\mathbf{z}_j$ are column vectors. We use $\mathcal{B}[J]$ to denote the leading subtensor $(b_{i_1 i_2 \ldots i_m})_{i_j \in J}$, where J is an arbitrary index set. Since $s > \rho(\mathcal{B}) \geq \rho(\mathcal{B}[I_t])$, the leading subtensors $s\mathcal{I} - \mathcal{B}[I_t]$ are irreducible nonsingular $\mathcal{M}$-tensors. Hence they are also semi-positive, that is, there exists $\mathbf{x}_t > \mathbf{0}$ with $s\mathbf{x}_t^{[m-1]} - \mathcal{B}[I_t]\mathbf{x}_t^{m-1} > 0$ for all $t = 1, 2, \ldots, k$.

Consider the leading subtensor $\mathcal{B}[I_1 \sqcup I_2]$ first. For all vectors $\mathbf{z}_1$ and $\mathbf{z}_2$ of lengths n_1 and $n_2 - n_1$, respectively, we have

$$\mathcal{B}[I_1 \sqcup I_2]\begin{bmatrix} \mathbf{z}_1 \\ \mathbf{z}_2 \end{bmatrix}^{m-1} = \begin{bmatrix} \mathcal{B}[I_1]\mathbf{z}_1^{m-1} + \displaystyle\sum_{\mathbf{a} \neq (1,1,\ldots,1)} \mathcal{B}_{(1,\mathbf{a})}\mathbf{z}_{\mathbf{a}}^{m-1} \\ \mathcal{B}[I_2]\mathbf{z}_2^{m-1} \end{bmatrix},$$

where $\mathbf{a}$ only contains 1 and 2. Take $\mathbf{z}_1 = \mathbf{x}_1$ and $\mathbf{z}_2 = \varepsilon\mathbf{x}_2$, where $\varepsilon \in (0,1)$ satisfies

$$\varepsilon \cdot \sum_{\mathbf{a} \neq (1,1,\ldots,1)} \mathcal{B}_{(1,\mathbf{a})}\mathbf{x}_{\mathbf{a}}^{m-1} < s\mathbf{x}_1^{[m-1]} - \mathcal{B}[I_1]\mathbf{x}_1^{m-1}.$$

Since $\displaystyle\sum_{\mathbf{a} \neq (1,1,\ldots,1)} \mathcal{B}_{(1,\mathbf{a})}\mathbf{z}_{\mathbf{a}}^{m-1} \leq \varepsilon\Big(\displaystyle\sum_{\mathbf{a} \neq (1,1,\ldots,1)} \mathcal{B}_{(1,\mathbf{a})}\mathbf{x}_{\mathbf{a}}^{m-1}\Big)$, it can be deduced that $\mathcal{B}[I_1]\mathbf{z}_1^{m-1} + \displaystyle\sum_{\mathbf{a} \neq (1,1,\ldots,1)} \mathcal{B}_{(1,\mathbf{a})}\mathbf{z}_{\mathbf{a}}^{m-1} < s\mathbf{z}_1^{[m-1]}$. Therefore we obtain

$$\begin{bmatrix} \mathbf{x}_1 \\ \varepsilon\mathbf{x}_2 \end{bmatrix} > \mathbf{0} \quad \text{and} \quad s\begin{bmatrix} \mathbf{x}_1 \\ \varepsilon\mathbf{x}_2 \end{bmatrix}^{[m-1]} - \mathcal{B}\begin{bmatrix} \mathbf{x}_1 \\ \varepsilon\mathbf{x}_2 \end{bmatrix}^{m-1} > \mathbf{0},$$

so $\mathcal{A}[I_1 \sqcup I_2] = s\mathcal{I} - \mathcal{B}[I_1 \sqcup I_2]$ is a semi-positive $\mathcal{Z}$-tensor.

Assume that $\mathcal{A}[I_1 \sqcup I_2 \sqcup \cdots \sqcup I_t]$ is a semi-positive $\mathcal{Z}$-tensor. Consider the leading subtensor $\mathcal{A}[I_1 \sqcup I_2 \sqcup \cdots \sqcup I_{t+1}]$ next. Substituting the index sets I_1 and I_2 above with $I_1 \sqcup I_2 \sqcup \cdots \sqcup I_t$ and I_{t+1}, respectively, we can conclude that $\mathcal{A}[I_1 \sqcup I_2 \sqcup \cdots \sqcup I_{t+1}]$ is also a semi-positive $\mathcal{Z}$-tensor. Thus by induction, we prove that the weakly reducible nonsingular $\mathcal{M}$-tensor $\mathcal{A}$ is a semi-positive $\mathcal{Z}$-tensor. $\qquad\square$

Combining Lemmas 5.17 and 5.18, we finish the proof of Theorem 5.7. Thus all the properties of semi-positive $\mathcal{Z}$-tensors we investigated above are shared with nonsingular $\mathcal{M}$-tensors. Therefore semi-positivity can be employed to study the nonsingular $\mathcal{M}$-tensors.

5.3.4 General $\mathcal{M}$-Tensors

We have discussed the nonsingular $\mathcal{M}$-tensors above; however, general $\mathcal{M}$-tensors are also useful. The examples can be found in the literature. For instance, the Laplacian tensor $\mathcal{L}$ of a hypergraph [59, 61, 62, 101] is an $\mathcal{M}$-tensor but is not nonsingular.

An $\mathcal{M}$-tensor can be written as $\mathcal{A} = s\mathcal{I} - \mathcal{B}$, where $\mathcal{B}$ is nonnegative and $s \geq \rho(\mathcal{B})$. It is easy to verify that the tensor $\mathcal{A}_\varepsilon = \mathcal{A} + \varepsilon\mathcal{I}$ $(\varepsilon > 0)$ is then a nonsingular $\mathcal{M}$-tensor and $\mathcal{A}$ is the limit of a sequence of $\mathcal{A}_\varepsilon$ when $\varepsilon \to 0$. Since all the diagonal entries of a semi-positive $\mathcal{Z}$-tensor, that is, a nonsingular $\mathcal{M}$-tensor, are positive, the diagonal entries of a general $\mathcal{M}$-tensor, as the limit of a sequence of nonsingular $\mathcal{M}$-tensor, must be nonnegative. Thus we have proved the following proposition.

Proposition 5.19. *An $\mathcal{M}$-tensor has nonnegative diagonal entries.*

The conception semi-positivity [9] can be extended as follows.

Definition 5.6 (Semi-nonnegative tensor). *We call $\mathcal{A}$ a semi-nonnegative tensor, if there exists $\mathbf{x} > \mathbf{0}$ such that $\mathcal{A}\mathbf{x}^{m-1} \geq \mathbf{0}$.*

Unlike the semi-positive case, being a semi-nonnegative $\mathcal{Z}$-tensor is not equivalent to being a general $\mathcal{M}$-tensor. Actually, a semi-nonnegative $\mathcal{Z}$-tensor must be an $\mathcal{M}$-tensor, but the converse is not true. The proof of the first statement is analogous to Lemma 5.17, so we have the next theorem.

Theorem 5.20. *A semi-nonnegative $\mathcal{Z}$-tensor is an $\mathcal{M}$-tensor.*

Conversely, we can give a counter example of an $\mathcal{M}$-tensor that is not semi-nonnegative. Let $\mathcal{B}$ be a nonnegative tensor of size $2 \times 2 \times 2 \times 2$ with the entries:

$$b_{1111} = 2, \ b_{1122} = b_{2222} = 1, \text{ and } b_{i_1i_2i_3i_4} = 0 \text{ otherwise.}$$

Then the spectral radius of $\mathcal{B}$ is apparently 2. Let $\mathcal{A} = 2\mathcal{I} - \mathcal{B}$, then $\mathcal{A}$ is an $\mathcal{M}$-tensor with entries

$$a_{1122} = -1, \ a_{2222} = 1, \text{ and } a_{i_1i_2i_3i_4} = 0 \text{ otherwise.}$$

Thus for every positive vector $\mathbf{x}$, the first component of $\mathcal{A}\mathbf{x}^3$ is always negative and the second one is positive, which means that there is no positive vector $\mathbf{x}$ such that $\mathcal{A}\mathbf{x}^3 \geq \mathbf{0}$. Therefore $\mathcal{A}$ is an $\mathcal{M}$-tensor but not semi-nonnegative. Of course, there exist some semi-nonnegative $\mathcal{M}$-tensors.

Proposition 5.21. *The following tensors are semi-nonnegative:*

(1) *A diagonally dominant $\mathcal{Z}$-tensor with nonnegative diagonal entries.*

(2) *A weakly irreducible $\mathcal{M}$-tensor.*

(3) *Consider the weakly reducible $\mathcal{M}$-tensor $\mathcal{A} = s\mathcal{I} - \mathcal{B}$, where $\mathcal{B} \geq 0$ and $s = \rho(\mathcal{B})$, and $I_1, I_2, \ldots, I_k$ are defined as in Lemma 5.18. Let*

$$\rho(\mathcal{B}[I_t]) \begin{cases} < s, & t = 1, 2, \ldots, k_1, \\ = s, & t = k_1 + 1, k_1 + 2, \ldots, k \end{cases}$$

and the entries of $\mathcal{B}[I_{k_1+1} \sqcup I_{k_1+2} \sqcup \cdots \sqcup I_k]$ be all zeros except the ones in the leading subtensors $\mathcal{B}[I_{k_1+1}], \mathcal{B}[I_{k_1+2}], \ldots, \mathcal{B}[I_k]$.

(4) *A symmetric $\mathcal{M}$-tensor.*

The proofs for (1)–(3) are similar to the semi-positive case and (4) is a direct consequence of (3).

5.4   Monotonicity

Following the condition (C9) in Section 5.1.2, we generalize *monotonicity* [9] from nonsingular M-matrices to higher order tensors. The monotonicity of a nonsingular M-matrix implies that its inverse matrix is nonnegative, which is interesting for some applications associated with nonnegative solutions of linear systems.

5.4.1   Definitions

Definition 5.7 (Monotone tensor). *We call $\mathcal{A}$ monotone, if $\mathcal{A}\mathbf{x}^{m-1} \geq \mathbf{0}$ implies $\mathbf{x} \geq \mathbf{0}$.*

It is easy to show that the set of all monotone tensors is not empty, since the even-order diagonal tensors with all positive diagonal entries belong to this set. However, an odd-order tensor is never a monotone tensor since $\mathcal{A}\mathbf{x}^{m-1} \geq \mathbf{0}$ implies $\mathcal{A}(-\mathbf{x})^{m-1} \geq \mathbf{0}$, thus we cannot guarantee that $\mathbf{x}$ is nonnegative.

Sometimes we use the following as an equivalent definition of monotone tensor for convenience.

Lemma 5.22. *An even-order tensor $\mathcal{A}$ is monotone if and only if $\mathcal{A}\mathbf{x}^{m-1} \leq \mathbf{0}$ implies $\mathbf{x} \leq \mathbf{0}$.*

Proof. Suppose that $\mathcal{A}$ is a monotone tensor. Since $\mathcal{A}\mathbf{x}^{m-1} \leq \mathbf{0}$ and $m-1$ is odd, we have
$$\mathcal{A}(-\mathbf{x})^{m-1} = -\mathcal{A}\mathbf{x}^{m-1} \geq \mathbf{0}.$$

Then $-\mathbf{x} \geq \mathbf{0}$ by the definition, thus $\mathbf{x} \leq \mathbf{0}$.

If $\mathcal{A}\mathbf{x}^{m-1} \leq \mathbf{0}$ implies $\mathbf{x} \leq \mathbf{0}$, then $\mathcal{A}\mathbf{y}^{m-1} \geq \mathbf{0}$ implies
$$\mathcal{A}(-\mathbf{y})^{m-1} = -\mathcal{A}\mathbf{y}^{m-1} \leq \mathbf{0}.$$

Therefore $-\mathbf{y} \leq \mathbf{0}$, which is equivalent to $\mathbf{y} \geq \mathbf{0}$. Thus $\mathcal{A}$ is a monotone tensor. □

5.4.2   Properties

We shall prove that a monotone $\mathcal{Z}$-tensor is also a nonsingular $\mathcal{M}$-tensor. Before that, we need some lemmas.

Lemma 5.23. *An even-order monotone tensor has no zero H-eigenvalue.*

Proof. Let $\mathcal{A}$ be an even-order monotone tensor. If $\mathcal{A}$ has a zero H-eigenvalue, that is, there is a nonzero $\mathbf{x} \in \mathbb{R}^n$ such that $\mathcal{A}\mathbf{x}^{m-1} = \mathbf{0}$, then $\mathcal{A}(\alpha\mathbf{x})^{m-1} = \mathbf{0}$ for all $\alpha \in \mathbb{R}$. Thus we cannot ensure that $\alpha\mathbf{x} \geq \mathbf{0}$, which contradicts the definition of a monotone tensor. Therefore $\mathcal{A}$ has no zero H-eigenvalue. □

Lemma 5.24. *Every H^+-eigenvalue of an even-order monotone tensor is non-negative.*

Proof. Let $\mathcal{A}$ be an even-order monotone tensor and λ be an H^+-eigenvalue of $\mathcal{A}$, that is, there is a nonzero $\mathbf{x} \in \mathbb{R}^n_+$ such that $\mathcal{A}\mathbf{x}^{m-1} = \lambda\mathbf{x}^{[m-1]}$. Then we have $\mathcal{A}(-\mathbf{x})^{m-1} = -\lambda\mathbf{x}^{[m-1]}$, since $m-1$ is odd. If $\lambda < 0$, then $\mathcal{A}(-\mathbf{x})^{m-1} = -\lambda\mathbf{x}^{[m-1]} \geq \mathbf{0}$, and $\mathbf{x} \leq \mathbf{0}$. It contradicts that $\mathbf{x}$ is nonzero and nonnegative, thus $\lambda \geq 0$. $\qquad\square$

The next theorem follows directly from Lemma 5.23 and Lemma 5.24.

Theorem 5.25. *Every H^+-eigenvalue of an even-order monotone tensor is positive.*

By applying this result, we can now reveal the relationship between the sets of even-order monotone $\mathcal{Z}$-tensors and nonsingular $\mathcal{M}$-tensors.

Theorem 5.26. *An even-order monotone $\mathcal{Z}$-tensor is also a nonsingular $\mathcal{M}$-tensor.*

Proof. Let $\mathcal{A}$ be an even-order monotone $\mathcal{Z}$-tensor. We have $\mathcal{A} = s\mathcal{I} - \mathcal{B}$, where $\mathcal{B} \geq 0$. Then $\rho(\mathcal{B})$ is an H^+-eigenvalue of $\mathcal{B}$ by Perron-Frobenius theorem for nonnegative tensors, that is, there is a nonzero $\mathbf{x} \geq \mathbf{0}$ with $\mathcal{B}\mathbf{x}^{m-1} = \rho(\mathcal{B})\mathbf{x}^{[m-1]}$. Then

$$\mathcal{A}\mathbf{x}^{m-1} = (s\mathcal{I} - \mathcal{B})\mathbf{x}^{m-1} = (s - \rho(\mathcal{B}))\mathbf{x}^{[m-1]},$$

or $s - \rho(\mathcal{B})$ is an H^+-eigenvalue of $\mathcal{A}$. From Theorem 5.25, the H^+-eigenvalue $s - \rho(\mathcal{B}) > 0$, thus $s > \rho(\mathcal{B})$. Therefore $\mathcal{A}$ is a nonsingular $\mathcal{M}$-tensor. $\qquad\square$

This theorem tells us that

{Monotone $\mathcal{Z}$-tensors of even order} $\subseteq$ {Nonsingular $\mathcal{M}$-tensors of even order}.

However, we show that not every nonsingular $\mathcal{M}$-tensor is monotone in the following subsection. The result for matrices is untrue when the order is larger than 2.

Next, we will present some properties of monotone $\mathcal{Z}$-tensors.

Proposition 5.27. *An even-order monotone $\mathcal{Z}$-tensor has positive diagonal entries.*

Proof. Let $\mathcal{A}$ be an even-order monotone $\mathcal{Z}$-tensor. Consider $\mathcal{A}\mathbf{e}_i^{m-1}$ ($i = 1, 2, \ldots, n$), where $\mathbf{e}_i$ denotes the vector with only one nonzero entry 1 in the i-th position. We have

$$(\mathcal{A}\mathbf{e}_i^{m-1})_i = a_{ii\ldots i}, \qquad i = 1, 2, \ldots, n$$
$$(\mathcal{A}\mathbf{e}_i^{m-1})_j = a_{ji\ldots i} \leq 0, \qquad j \neq i.$$

If $a_{ii\ldots i} \leq 0$, then $\mathcal{A}\mathbf{e}_i^{m-1} \leq \mathbf{0}$, which indicates $\mathbf{e}_i \leq \mathbf{0}$ by Lemma 5.22. By contradiction, we have $a_{ii\ldots i} > 0$ for $i = 1, 2, \ldots, n$. $\qquad\square$

The next proposition shows that some row of a monotone $\mathcal{Z}$-tensor is strictly diagonally dominant.

Proposition 5.28. *Let $\mathcal{A}$ be an even-order monotone $\mathcal{Z}$-tensor. Then*

$$a_{ii\ldots i} > \sum_{(i_2,i_3,\ldots,i_m)\neq(i,i,\ldots,i)} |a_{ii_2\ldots i_m}|$$

for some $i \in \{1,2,\ldots,n\}$.

Proof. Consider $\mathcal{A}e^{m-1}$, where $\mathbf{e}$ denotes the all-one vector. We have

$$(\mathcal{A}e^{m-1})_i = a_{ii\ldots i} + \sum_{(i_2,i_3,\ldots,i_m)\neq(i,i,\ldots,i)} a_{ii_2\ldots i_m} = a_{ii\ldots i} - \sum_{(i_2,i_3,\ldots,i_m)\neq(i,i,\ldots,i)} |a_{ii_2\ldots i_m}|,$$

since $a_{ii_2\ldots i_m} \leq 0$ for $(i_2,i_3,\ldots,i_m) \neq (i,i,\ldots,i)$.

If $a_{ii\ldots i} \leq \sum_{(i_2,i_3,\ldots,i_m)\neq(i,i,\ldots,i)} |a_{ii_2\ldots i_m}|$ for all $i = 1,2,\ldots,n$, then $\mathcal{A}e^{m-1} \leq \mathbf{0}$,

which indicates $\mathbf{e} \leq \mathbf{0}$ by Lemma 5.22. Therefore $a_{ii\ldots i} > \sum_{(i_2,i_3,\ldots,i_m)\neq(i,i,\ldots,i)} |a_{ii_2\ldots i_m}|$

for some i. $\qquad\square$

Proposition 5.29. *Let $\mathcal{A}$ be an even-order monotone $\mathcal{Z}$-tensor. Then $\mathcal{A} + \mathcal{D}$ has positive H^+-eigenvalues for each nonnegative diagonal tensor $\mathcal{D}$.*

Proof. If $\mathcal{A} + \mathcal{D}$ has a nonpositive H^+-eigenvalue, then there exists a nonzero vector $\mathbf{x} \geq \mathbf{0}$ such that $(\mathcal{A} + \mathcal{D})\mathbf{x}^{m-1} = \lambda\mathbf{x}^{[m-1]}$ and $\lambda \leq 0$. Consequently, we have $\mathcal{A}\mathbf{x}^{m-1} = \lambda\mathbf{x}^{[m-1]} - \mathcal{D}\mathbf{x}^{m-1} \leq \mathbf{0}$, since $\mathbf{x}$ and $\mathcal{D}$ are nonnegative and λ is nonpositive, which leads to the contradiction $\mathbf{x} \leq \mathbf{0}$ from the definition of monotone $\mathcal{Z}$-tensors. Therefore $\mathcal{A} + \mathcal{D}$ has no nonpositive H^+-eigenvalues for all nonnegative diagonal tensor $\mathcal{D}$. $\qquad\square$

Furthermore, the monotone $\mathcal{Z}$-tensor also has the following two equivalent characterizations, following conditions (C10) and (C11).

Proposition 5.30. *A $\mathcal{Z}$-tensor $\mathcal{A}$ is monotone if and only if there exists a positive diagonal tensor $\mathcal{D}$ and a monotone $\mathcal{Z}$-tensor $\mathcal{C}$ such that $\mathcal{A} = \mathcal{D}\mathcal{C}$.*

Proof. Let $\mathcal{D}$ be the diagonal part of $\mathcal{A}$ and $\mathcal{C} = \mathcal{D}^{-1}\mathcal{A}$. Clearly, $\mathcal{A} = \mathcal{D}\mathcal{C}$.

If $\mathcal{A}$ is a monotone $\mathcal{Z}$-tensor, then $\mathcal{D}$ is positive diagonal and $\mathcal{A}\mathbf{x}^{m-1} \geq \mathbf{0}$ implies $\mathbf{x} \geq \mathbf{0}$. When $\mathcal{C}\mathbf{x}^{m-1} \geq \mathbf{0}$, the vector $\mathcal{A}\mathbf{x}^{m-1} = \mathcal{D}\left((\mathcal{C}\mathbf{x}^{m-1})^{[\frac{1}{m-1}]}\right)^{m-1}$ is also nonnegative, thus $\mathbf{x} \geq \mathbf{0}$. Since $\mathcal{C}$ is also a $\mathcal{Z}$-tensor, it is a monotone $\mathcal{Z}$-tensor.

If $\mathcal{C}$ is a monotone $\mathcal{Z}$-tensor and $\mathcal{D}$ is positive diagonal, then $\mathcal{C}\mathbf{x}^{m-1} \geq \mathbf{0}$ implies $\mathbf{x} \geq \mathbf{0}$. When $\mathcal{A}\mathbf{x}^{m-1} \geq \mathbf{0}$, the vector $\mathcal{C}\mathbf{x}^{m-1} = \mathcal{D}^{-1}\left((\mathcal{A}\mathbf{x}^{m-1})^{[\frac{1}{m-1}]}\right)^{m-1}$ is also nonnegative, thus $\mathbf{x} \geq \mathbf{0}$. Therefore $\mathcal{A}$ is a monotone $\mathcal{Z}$-tensor. $\qquad\square$

Proposition 5.31. *A $\mathcal{Z}$-tensor $\mathcal{A}$ is monotone if and only if there exists a positive diagonal tensor $\mathcal{D}$ and a nonnegative tensor $\mathcal{E}$ such that $\mathcal{A} = \mathcal{D} - \mathcal{E}$ and $(\mathcal{D}^{-1}\mathcal{E})\mathbf{x}^{m-1} \leq \mathbf{x}^{[m-1]}$ implies $\mathbf{x} \geq \mathbf{0}$.*

Proof. Let $\mathcal{D}$ be the diagonal tensor of $\mathcal{A}$ and $\mathcal{E} = \mathcal{D} - \mathcal{A}$. Clearly, $\mathcal{A} = \mathcal{D} - \mathcal{E}$ and $\mathcal{D}^{-1}\mathcal{E} = \mathcal{I} - \mathcal{D}^{-1}\mathcal{A}$.

If $\mathcal{A}$ is a monotone $\mathcal{Z}$-tensor, then $\mathcal{D}$ is positive diagonal and $\mathcal{A}\mathbf{x}^{m-1} \geq \mathbf{0}$ implies $\mathbf{x} \geq \mathbf{0}$. When $(\mathcal{D}^{-1}\mathcal{E})\mathbf{x}^{m-1} \leq \mathbf{x}^{[m-1]}$, we have $\mathcal{E}\mathbf{x}^{m-1} \leq \mathcal{D}\mathbf{x}^{m-1}$, and thus $\mathcal{A}\mathbf{x}^{m-1} \geq \mathbf{0}$. This indicates $\mathbf{x} \geq \mathbf{0}$.

Assume that $(\mathcal{D}^{-1}\mathcal{E})\mathbf{x}^{m-1} \leq \mathbf{x}^{[m-1]}$ implies $\mathbf{x} \geq \mathbf{0}$. When $\mathcal{A}\mathbf{x}^{m-1} \geq \mathbf{0}$, we have $\mathcal{D}\mathbf{x}^{m-1} \geq \mathcal{E}\mathbf{x}^{m-1}$, and thus $(\mathcal{D}^{-1}\mathcal{E})\mathbf{x}^{m-1} \leq \mathbf{x}^{[m-1]}$, implying $\mathbf{x} \geq \mathbf{0}$. Since $\mathcal{A}$ is also a $\mathcal{Z}$-tensor, it is a monotone $\mathcal{Z}$-tensor. $\qquad\square$

5.4.3 A Counter Example

We will give a 4th-order counter example in this section to show that the set of all monotone $\mathcal{Z}$-tensor is a proper subset of the set of all nonsingular $\mathcal{M}$-tensor, when the order is larger than 2.

Recall the outer product $X \circ Y$, where $(X \circ Y)_{i_1 i_2 i_3 i_4} = x_{i_1 i_2} \cdot y_{i_3 i_4}$. Let $\mathcal{J} = I_n \circ I_n$, where I_n denotes the $n \times n$ identity matrix. It is obvious that the spectral radius $\rho(\mathcal{J}) = n$, since the sum of each row of $\mathcal{J}$ equals n. Take

$$\mathcal{A} = s\mathcal{I} - \mathcal{J} \ (s > n), \quad \mathbf{x} = \begin{bmatrix} 1 \\ \vdots \\ 1 \\ -\delta \end{bmatrix} \ (0 < \delta < 1).$$

Then $\mathcal{A}$ is a nonsingular $\mathcal{M}$-tensor and

$$\mathcal{A}\mathbf{x}^3 = s\mathbf{x}^{[3]} - (\mathbf{x}^\top\mathbf{x})\mathbf{x} = s \begin{bmatrix} 1 \\ \vdots \\ 1 \\ -\delta^3 \end{bmatrix} - (n-1+\delta^2) \begin{bmatrix} 1 \\ \vdots \\ 1 \\ -\delta \end{bmatrix} = \begin{bmatrix} s - n + 1 - \delta^2 \\ \vdots \\ s - n + 1 - \delta^2 \\ (n - 1 + (1 - s)\delta^2)\delta \end{bmatrix}.$$

When $\delta \leq \sqrt{\frac{n-1}{s-1}}$, the vector $\mathcal{A}\mathbf{x}^3$ is nonnegative while $\mathbf{x}$ is not. Therefore $\mathcal{A}$ is not a monotone $\mathcal{Z}$-tensor, although it is a nonsingular $\mathcal{M}$-tensor.

5.4.4 A Nontrivial Monotone $\mathcal{Z}$-Tensor

Let

$$\mathcal{B} = \mathbf{a}^{[2k-1]} \circ \underbrace{\mathbf{b} \circ \mathbf{b} \circ \cdots \circ \mathbf{b}}_{2k-1}$$

be a tensor of order $2k$, where $\mathbf{a}$ and $\mathbf{b}$ are nonnegative vectors. It is easy to compute that $\rho(\mathcal{B}) = (\mathbf{b}^\top\mathbf{a})^{2k-1}$. Then $\mathcal{A} = s\mathcal{I} - \mathcal{B} \ (s > (\mathbf{b}^\top\mathbf{a})^{2k-1})$ is a nonsingular $\mathcal{M}$-tensor. For each $\mathbf{x}$, we get

$$\mathcal{A}\mathbf{x}^{2k-1} = s\mathbf{x}^{[2k-1]} - \mathbf{a}^{[2k-1]}(\mathbf{b}^\top\mathbf{x})^{2k-1}.$$

When $\mathcal{A}\mathbf{x}^{2k-1} \geq 0$, we have

$$s^{\frac{1}{2k-1}} \cdot \mathbf{x} \geq \mathbf{a} \cdot (\mathbf{b}^\top\mathbf{x}),$$

thus

$$s^{\frac{1}{2k-1}} \cdot (\mathbf{b}^\top\mathbf{x}) \geq (\mathbf{b}^\top\mathbf{a})(\mathbf{b}^\top\mathbf{x}).$$

Since $s > (\mathbf{b}^\top\mathbf{a})^{2k-1}$, we can conclude $\mathbf{b}^\top\mathbf{x} \geq 0$. Therefore $\mathbf{x} \geq \mathbf{a} \cdot \frac{(\mathbf{b}^\top\mathbf{x})}{s} \geq 0$, which indicates that $\mathcal{A}$ is a monotone $\mathcal{Z}$-tensor.

5.5 An Extension of $\mathcal{M}$-Tensors

Inspired by the conception of H-matrix [9], we can further extend $\mathcal{M}$-tensors to $\mathcal{H}$-tensors. More details can be found in [70, 104, 122, 124]. First, we need a concept called the comparison tensor.

Definition 5.8. *Let $\mathcal{A} = (a_{i_1 i_2 \ldots i_m})$ be a tensor of order m and dimension n. We call $\mathcal{M}(\mathcal{A}) = (m_{i_1 i_2 \ldots i_m})$ the* comparison tensor *of $\mathcal{A}$ if*

$$m_{i_1 i_2 \ldots i_m} = \begin{cases} +|a_{i_1 i_2 \ldots i_m}|, & \text{if } (i_2, i_3, \ldots, i_m) = (i_1, i_1, \ldots, i_1), \\ -|a_{i_1 i_2 \ldots i_m}|, & \text{if } (i_2, i_3, \ldots, i_m) \neq (i_1, i_1, \ldots, i_1). \end{cases}$$

Then we can describe an $\mathcal{H}$-tensor.

Definition 5.9. *We call a tensor a* (nonsingular) $\mathcal{H}$-tensor, *if its comparison tensor is a (nonsingular) $\mathcal{M}$-tensor.*

Nonsingular $\mathcal{H}$-tensors have a property called quasi-strictly diagonally dominant, which can be proved directly from the properties of nonsingular $\mathcal{M}$-tensors. We leave the proof to the reader.

Theorem 5.32. *A tensor $\mathcal{A}$ is a nonsingular $\mathcal{H}$-tensor if and only if it is quasi-strictly diagonally dominant, that is, there exist n positive real numbers $d_1, d_2, \ldots, d_n$ such that*

$$|a_{ii \ldots i}| d_i^{m-1} > \sum_{(i_2, i_3, \ldots, i_m) \neq (i, i, \ldots, i)} |a_{i i_2 \ldots i_m}| d_{i_2} \ldots d_{i_m}, \quad i = 1, 2, \ldots, n.$$

Similarly to the nonsingular $\mathcal{M}$-tensor, a nonsingular $\mathcal{H}$-tensor has other characteristics.

Theorem 5.33. *The following conditions are equivalent:*

(E1) *A tensor $\mathcal{A}$ is a nonsingular $\mathcal{H}$-tensor.*

(E2) *There exists a positive diagonal matrix D such that $\mathcal{A}D^{m-1}$ is strictly diagonally dominant.*

(E3) *There exist two positive diagonal matrix D_1 and D_2 such that $D_1 \mathcal{A} D_2^{m-1}$ is strictly diagonally dominant.*

Employing these equivalent definitions, we have immediately the next proposition.

Proposition 5.34. *If $\mathcal{A}$ is a nonsingular $\mathcal{H}$-tensor, then there is some index i such that*

$$|a_{ii \ldots i}| > \sum_{(i_2, i_3, \ldots, i_m) \neq (i, i, \ldots, i)} |a_{i i_2 \ldots i_m}|.$$

Proof. Otherwise, let $\mathbf{d}$ be an arbitrary positive vector and d_i be one of the minimum components of $\mathbf{d}$. Then

$$|a_{ii \ldots i}| d_i^{m-1} \leq \sum_{(i_2, i_3, \ldots, i_m) \neq (i, i, \ldots, i)} |a_{i i_2 \ldots i_m}| d_i^{m-1}$$

$$\leq \sum_{(i_2, i_3, \ldots, i_m) \neq (i, i, \ldots, i)} |a_{i i_2 \ldots i_m}| d_{i_2} \ldots d_{i_m},$$

which contradicts Theorem 5.32. $\qquad\qquad\qquad\qquad\qquad\qquad\qquad\qquad \square$

We have proven in the previous section that for a nonsingular $\mathcal{M}$-tensor both the diagonal entries and the real parts of the eigenvalues are positive. This is not a coincidence. We shall show shortly that the number of positive or negative eigenvalues of a nonsingular $\mathcal{H}$-tensor is dependent on the number of positive or negative diagonal entries, respectively. Before that, recall the Gershgorin circle theorem for tensors.

Theorem 5.35 ([98]). *Let $\mathcal{A}$ be an mth-order n-dimensional tensor.*

1. *The eigenvalues of $\mathcal{A}$ lie in the union of n disks*

$$G_i(\mathcal{A}) = \left\{ z \in \mathbb{C} : \left| z - a_{ii\ldots i} \right| \leq \sum_{(i_2, i_3, \ldots, i_m) \neq (i, i, \ldots, i)} \left| a_{ii_2 \ldots i_m} \right| \right\}, \quad i = 1, 2, \ldots, n.$$

2. *If k of these n disks are disjoint with the remaining $n - k$ disks, then there are exactly $k(m-1)^{n-1}$ eigenvalues which lie in the union of these k disks.*

We have the following depiction of the real parts of the eigenvalues of a nonsingular $\mathcal{H}$-tensor.

Theorem 5.36. *Let $\mathcal{A}$ be an mth-order n-dimensional tensor. If $\mathcal{A}$ has p positive and q negative diagonal entries ($p + q = n$), then it has $p(m-1)^{n-1}$ eigenvalues with positive real parts and $q(m-1)^{n-1}$ eigenvalues with negative real parts.*

Proof. By Theorem 5.33, there exists a positive diagonal matrix D such that $\mathcal{A}D^{m-1}$ is strictly diagonally dominant. Apparently, the tensor $D^{-(m-1)}\mathcal{A}D^{m-1}$ is also strictly diagonally dominant, and it shares the same diagonal entries and eigenvalues with $\mathcal{A}$. For a strictly diagonally dominant tensor, if it has p positive and q negative diagonal entries, then there must be p Gershgorin circles on the right half plane and q Gershgorin circles on the left half plane. Thus we obtain the final results. $\qquad\square$

5.6 Summation

In this chapter, we give the proofs or the counter examples to show these three relations between different sets of tensors:

$$\{\text{semi-positive } \mathcal{Z}\text{-tensors}\} = \{\text{nonsingular } \mathcal{M}\text{-tensors}\},$$
$$\{\text{semi-nonnegative } \mathcal{Z}\text{-tensors}\} \subsetneqq \{\text{general } \mathcal{M}\text{-tensors}\},$$
$$\{\text{even-order monotone } \mathcal{Z}\text{-tensors}\} \subsetneqq \{\text{even-order nonsingular } \mathcal{M}\text{-tensors}\}.$$

We list the equivalent definitions of nonsingular $\mathcal{M}$-tensors involved in Chapters 5 and 6 as follows:
If $\mathcal{A}$ is a $\mathcal{Z}$-tensor, then the following conditions are equivalent:

(D1) $\mathcal{A}$ is a nonsingular $\mathcal{M}$-tensor;

(D2) The real part of each eigenvalue of $\mathcal{A}$ is positive.

(D3) Every real eigenvalue of $\mathcal{A}$ is positive.

(D4) Every H-eigenvalue of $\mathcal{A}$ is positive.

(D5) Every H^+-eigenvalue of $\mathcal{A}$ is positive.

(D6) $\mathcal{A}$ is semi-positive; that is, there exists $\mathbf{x} > \mathbf{0}$ with $\mathcal{A}\mathbf{x}^{m-1} > \mathbf{0}$.

(D7) There exists $\mathbf{x} \geq \mathbf{0}$ with $\mathcal{A}\mathbf{x}^{m-1} > \mathbf{0}$.

(D8) $\mathcal{A}$ has all positive diagonal entries, and there exists a positive diagonal matrix D such that $\mathcal{A}D^{m-1}$ is strictly diagonally dominant.

(D9) $\mathcal{A}$ has all positive diagonal entries, and there exist two positive diagonal matrices D_1 and D_2 such that $D_1\mathcal{A}D_2^{m-1}$ is strictly diagonally dominant.

(D10) There exists a positive diagonal tensor $\mathcal{D}$ and a nonsingular $\mathcal{M}$-tensor $\mathcal{C}$ with $\mathcal{A} = \mathcal{D}\mathcal{C}$.

(D11) There exists a positive diagonal tensor $\mathcal{D}$ and a nonnegative tensor $\mathcal{E}$ such that $\mathcal{A} = \mathcal{D} - \mathcal{E}$ and there exists $\mathbf{x} > \mathbf{0}$ with $(\mathcal{D}^{-1}\mathcal{E})\mathbf{x}^{m-1} < \mathbf{x}^{[m-1]}$.

(D12) $\mathcal{A}\mathbf{x}^{m-1} = \mathbf{b}$ has a unique positive solution for every positive $\mathbf{b}$.

(D13) $\mathcal{A}\mathbf{x}^{m-1} - \mathcal{B}_{m-1}\mathbf{x}^{m-2} - \cdots - \mathcal{B}_2\mathbf{x} = \mathbf{b}$ has a unique positive solution for every positive $\mathbf{b}$ and nonnegative $\mathcal{B}_p$ $(p = 2, 3, \ldots, m-1)$.

(D14) $\mathcal{A}$ does not reverse the sign of any vector; that is, if $\mathbf{x} \neq \mathbf{0}$ and $\mathbf{b} = \mathcal{A}\mathbf{x}^{m-1}$, then $x_i^{m-1}b_i > 0$ for some index i.

Chapter 6

Multilinear Systems with $\mathcal{M}$-Tensors

A great disappointment in Chapter 5 is that monotonicity is not suitable for higher-order $\mathcal{M}$-tensors. Nevertheless, we can generalize this property, related to the nonnegative inverse property of $\mathcal{M}$-tensors, in a computational point of view. We now begin with multilinear systems of equations whose coefficient tensors are nonsingular $\mathcal{M}$-tensors.

6.1   Motivations

Solving multilinear systems is always an important issue in engineering and scientific computing [74]. The systems of multilinear equations can be expressed by tensor-vector products, just as we rewrite linear systems by matrix-vector products. Let $\mathcal{A}$ be an mth-order tensor in $\mathbb{C}^{n \times n \times \cdots \times n}$ and $\mathbf{b}$ a vector in $\mathbb{C}^n$. A *multilinear equation* can be expressed as

$$\mathcal{A}\mathbf{x}^{m-1} = \mathbf{b}.$$

Take an equation with the coefficient tensor $\mathcal{A} \in \mathbb{R}^{2 \times 2 \times 2}$ as an example. Equation $\mathcal{A}\mathbf{x}^2 = \mathbf{b}$ is a condensed form of

$$\begin{cases} a_{111}x_1^2 + (a_{112} + a_{121})x_1 x_2 + a_{122}x_2^2 = b_1, \\ a_{211}x_1^2 + (a_{212} + a_{221})x_1 x_2 + a_{222}x_2^2 = b_2. \end{cases}$$

We want to find x_1 and x_2 that satisfy the above two equations.

Denote the *solution set* of the multilinear equation $\mathcal{A}\mathbf{x}^{m-1} = \mathbf{b}$ by

$$\mathcal{A}^{-1}\mathbf{b} := \{\mathbf{x} \in \mathbb{C}^n : \mathcal{A}\mathbf{x}^{m-1} = \mathbf{b}\}.$$

Note that $\mathcal{A}^{-1}\mathbf{b}$ denotes a set rather than one of the solutions, and is generally different from the concept of "tensor inverse" in [15]. Furthermore, when $\mathcal{A}$ and $\mathbf{b}$ are both real, we define the *real*, the *nonnegative*, and the *positive solution sets* as

$$(\mathcal{A}^{-1}\mathbf{b})_{\mathbb{R}} := \{\mathbf{x} \in \mathbb{R}^n : \mathcal{A}\mathbf{x}^{m-1} = \mathbf{b}\},$$

$$(\mathcal{A}^{-1}\mathbf{b})_{+} := \{\mathbf{x} \in \mathbb{R}^n_{+} : \mathcal{A}\mathbf{x}^{m-1} = \mathbf{b}\}, \text{ and}$$

$$(\mathcal{A}^{-1}\mathbf{b})_{++} := \{\mathbf{x} \in \mathbb{R}^n_{++} : \mathcal{A}\mathbf{x}^{m-1} = \mathbf{b}\},$$

Theory and Computation of Tensors.
http://dx.doi.org/10.1016/B978-0-12-803953-3.50006-X
Copyright © 2016 Elsevier Ltd. All rights reserved.

respectively.

When are these sets nonempty and finite? When is there a unique element in these sets? How can we compute the solutions? Hilbert's Nullstellensatz[1] states that the solution set $\mathcal{A}^{-1}\mathbf{b}$ is nonempty if and only if there is no contradicting equation. To our best knowledge, there is no general answer to the other questions. Therefore we shall focus on some special equations with structured tensors.

One of the applications of the multilinear systems with $\mathcal{M}$-tensors is the sparsest solutions to tensor complementarity problems [78]:

$$(6.1) \qquad \min \|\mathbf{x}\|_0, \quad \text{s.t. } \mathcal{A}\mathbf{x}^{m-1} - \mathbf{b} \geq 0, \ \mathbf{x} \geq 0, \ \mathbf{x}^\top(\mathcal{A}\mathbf{x}^{m-1} - \mathbf{b}) = 0,$$

where $\|\mathbf{x}\|_0$ denotes the number of nonzero entries of $\mathbf{x}$.

This is generally an NP-hard problem. Luo et al. [78] suggested that if the tensor $\mathcal{A}$ is a $\mathcal{Z}$-tensor, then the sparsest solution of the above tensor complementarity problem can be achieved by solving the following polynomial programming problem

$$(6.2) \qquad\qquad \min \|\mathbf{x}\|_1, \quad \text{s.t. } \mathcal{A}\mathbf{x}^{m-1} = \mathbf{b}, \ \mathbf{x} \geq 0.$$

By using Theorem 6.3 in this chapter, Luo et al. proved that if $\mathcal{A}$ is a nonsingular $\mathcal{M}$-tensor, then the problem (6.2) is uniquely solvable and the solution is also an optimal solution to the problem (6.1). Furthermore, the problem (6.2) can be solved by our algorithm for $\mathcal{M}$-equations. Thus the NP-hard problem (6.1) can be solved in polynomial time.

Another application is the numerical solution of the partial differential equation with Dirichlet's boundary condition

$$(6.3) \qquad \begin{cases} u(\mathbf{x})^{m-2} \cdot \Delta u(\mathbf{x}) = -f(\mathbf{x}) \text{ in } \Omega, \\ u(\mathbf{x}) = g(\mathbf{x}) \text{ on } \partial\Omega, \end{cases} \quad (m = 3, 4, \dots)$$

where $\Delta = \sum_{k=1}^{d} \frac{\partial^2}{\partial x_k^2}$ and $\Omega = [0,1]^d$.

When $f(\cdot)$ is a constant function, this PDE is a nonlinear Klein-Gordon equation [82, 136]. As we know, the elliptic problem $\Delta u(\mathbf{x}) = -f(\mathbf{x})$ with Dirichlet's boundary condition $u(\mathbf{x}) = g(\mathbf{x})$ can be discretized into a linear system of equations whose coefficient matrix is a nonsingular M-matrix $L_h^{(d)} = \sum_{k=0}^{d-1} \underbrace{I \otimes \cdots \otimes I}_{k} \otimes L_h \otimes \underbrace{I \otimes \cdots \otimes I}_{d-k-1}$, where $h = 1/(n-1)$ and

$$L_h = \frac{1}{h^2} \begin{bmatrix} 1 & & & & & \\ -1 & 2 & -1 & & & \\ & -1 & 2 & -1 & & \\ & & \ddots & \ddots & \ddots & \\ & & & -1 & 2 & -1 \\ & & & & & 1 \end{bmatrix} \in \mathbb{R}^{n \times n}.$$

Similarly, the operator $u \mapsto u^{m-2} \cdot \Delta u$ can be discretized into an mth-order nonsingular $\mathcal{M}$-tensor $\mathcal{L}_h^{(d)} = \sum_{k=0}^{d-1} \underbrace{\mathcal{I} \otimes \cdots \otimes \mathcal{I}}_{k} \otimes \mathcal{L}_h \otimes \underbrace{\mathcal{I} \otimes \cdots \otimes \mathcal{I}}_{d-k-1}$, which

---

[1]German for "theorem of zeros", it can be found at http://en.wikipedia.org/wiki/Hilbert's_Nullstellensatz

satisfies $(\mathcal{L}_h \mathbf{u}^{m-1})_i = u_i^{m-2} \cdot (L_h \mathbf{u})_i$ for $i = 1, 2, \ldots, n$, where $\mathcal{L}_h$ is an mth-order $\mathcal{M}$-tensor with

$$
\begin{cases}
(\mathcal{L}_h)_{1,1,\ldots,1} = (\mathcal{L}_h)_{n,n,\ldots,n} = 1/h^2, \\
(\mathcal{L}_h)_{i,i,\ldots,i} = 2/h^2, \ i = 2, 3, \ldots, n-1, \\
(\mathcal{L}_h)_{i,i-1,i,\ldots,i} = (\mathcal{L}_h)_{i,i,i-1,\ldots,i} = \cdots = (\mathcal{L}_h)_{i,i,i,\ldots,i-1} = -1/h^2(m-1), \\
\qquad\qquad\qquad\qquad\qquad\qquad\qquad i = 2, 3, \ldots, n-1, \\
(\mathcal{L}_h)_{i,i+1,i,\ldots,i} = (\mathcal{L}_h)_{i,i,i+1,\ldots,i} = \cdots = (\mathcal{L}_h)_{i,i,i,\ldots,i+1} = -1/h^2(m-1), \\
\qquad\qquad\qquad\qquad\qquad\qquad\qquad i = 2, 3, \ldots, n-1.
\end{cases}
$$

The PDE in (6.3) is discretized into an $\mathcal{M}$-equation $\mathcal{L}_h^{(d)} \mathbf{u}^{m-1} = \mathbf{f}$. This class of multilinear equations can be regarded as a higher-order generalization of the one discussed in [69, 117]. We present the convergence analysis as an appendix.

Very recently, Hu and Qi [60] employed the existence and uniqueness theorem (Theorem 6.7) in this chapter to study a necessary and sufficient condition for the existence of a positive Perron vector.

6.2 Triangular Equations

A general multilinear equation is complicated to solve or analyze. A usual way to solve a system of polynomial equations is to compute its *triangular decomposition* [3, 4, 22], that is, to find a set of triangular systems defined later such that a point is a solution of the original system if and only if it is a solution of one of the triangular systems. Triangular systems are the base of the computations of polynomial systems. Thus we discuss the multilinear triangular equations briefly here.

The *diagonal* of a tensor $\mathcal{A}$ contains the entries $a_{ii\ldots i}$ with $i = 1, 2, \ldots, n$, and the other entries are called *off-diagonal*. A tensor is called *diagonal* if all its off-diagonal entries are zeros. A *diagonal equation* refers

$$
\mathcal{D}\mathbf{x}^{m-1} = \mathbf{b},
$$

where the coefficient m^{th}-order tensor $\mathcal{D}$ is diagonal.

If there is a zero on the diagonal of $\mathcal{D}$, then the solution set of this diagonal equation is either empty or infinite depending on whether the corresponding entry of $\mathbf{b}$ is zero. Next we consider $\mathcal{D}$ with nonzero diagonal entries. Thus it is easy to understand that the solution set $\mathcal{D}^{-1}\mathbf{b}$ has $(m-1)^n$ elements. When m is even, the real solution set $(\mathcal{D}^{-1}\mathbf{b})_{\mathbb{R}}$ has a unique element $\mathbf{x}$ with $x_i = (b_i/d_{ii\ldots i})^{1/(m-1)}$. Moreover, when $\mathcal{D}$ and the vector $\mathbf{b}$ are positive, the solution above is also the unique element in $(\mathcal{D}^{-1}\mathbf{b})_{++}$. When m is odd, the real solution set $(\mathcal{D}^{-1}\mathbf{b})_{\mathbb{R}}$ has at most 2^n elements $\mathbf{x}$ with $x_i = \pm(b_i/d_{ii\ldots i})^{1/(m-1)}$ if $d_{ii\ldots i}b_i \geq 0$ for all i, or else the real solution set is empty. Further when the diagonal of $\mathcal{D}$ and the vector $\mathbf{b}$ are positive, the positive solution set $(\mathcal{D}^{-1}\mathbf{b})_{++}$ has a unique element $\mathbf{x}$ with $x_i = (b_i/d_{ii\ldots i})^{1/(m-1)}$.

We can also define the triangular part of a tensor corresponding to triangular system of multilinear equations [3, 4, 22], which is different from those defined in [58, Section 5].

The *lower triangular part* of a tensor $\mathcal{A}$ contains the entries $a_{i_1 i_2 \ldots i_m}$ with $i_1 = 1, 2, \ldots, n$ and $i_2, \ldots, i_m \leq i_1$, and the other entries are in the *off-lower triangular part*. The *strictly lower part* consists of the entries $a_{i_1 i_2 \ldots i_m}$ with $i_1 = 1, 2, \ldots, n$ and $i_2, \ldots, i_m < i_1$. A tensor is called *lower triangular* if all entries not in the lower triangular part are zeros. A *lower triangular equation* refers

$$\mathcal{L}\mathbf{x}^{m-1} = \mathbf{b},$$

where the coefficient tensor $\mathcal{L}$ is lower triangular.

Similarly, the *upper triangular part* of a tensor $\mathcal{A}$ contains the entries $a_{i_1 i_2 \ldots i_m}$ with $i_1 = 1, 2, \ldots, n$ and $i_2, \ldots, i_m \geq i_1$, and the other entries are in the *off-upper triangular part*. The *strictly upper part* consists of the entries $a_{i_1 i_2 \ldots i_m}$ with $i_1 = 1, 2, \ldots, n$ and $i_2, \ldots, i_m > i_1$. A tensor is called *upper triangular* if all entries not in the upper triangular part are zeros. An *upper triangular equation* is referred to

$$\mathcal{U}\mathbf{x}^{m-1} = \mathbf{b},$$

where the coefficient tensor $\mathcal{U}$ is upper triangular.

We consider the lower triangular tensor with all diagonal entries being nonzero. As in the matrix case, a nonsingular lower triangular equation can be solved by forward substitution (refer to [49, Chapter 3] for the matrix version).

Algorithm 6.1 (Forward Substitution). *If $\mathcal{L} \in \mathbb{C}^{n \times n \times \cdots \times n}$ is lower triangular and $\mathbf{b} \in \mathbb{C}^n$, then this algorithm overwrites $\mathbf{b}$ with one of the solutions to $\mathcal{L}\mathbf{x}^{m-1} = \mathbf{b}$.*

$b_1 \leftarrow$ one of the $(m-1)$-st roots of $b_1/l_{11\ldots 1}$
for $i = 2 : n$
   for $k = 1 : m$
$$p_k \leftarrow \sum \left\{ l_{i i_2 \ldots i_m} \cdot \prod_{\substack{l=2,\ldots,m \\ l \neq p_1, \ldots, p_{k-1}}} b_{i_l} : i_2, \ldots, i_m \leq i, \right.$$
$$\left. i_{p_1}, \ldots, i_{p_{k-1}} \text{ are the only } k-1 \text{ indices equaling } i \right\}$$
   end
   $b_i \leftarrow$ one of the roots of $p_1 + p_2 t + \cdots + p_m t^{m-1} = b_i$
end

Applying the algorithm, we can analyze the existence and uniqueness of solutions. According to the fundamental theorem of algebra, the solution set $\mathcal{L}^{-1}\mathbf{b}$ has $(m-1)^n$ elements (counting multiplicity). When m is even, the polynomial equation of odd degree $m-1$ in the algorithm has at least one real root, so that the real solution set $(\mathcal{L}^{-1}\mathbf{b})_{\mathbb{R}}$ is nonempty. If this polynomial equation has a unique real root in each step, then the real solution set $(\mathcal{L}^{-1}\mathbf{b})_{\mathbb{R}}$ has a unique element. When m is odd, the existence of a real solution cannot be guaranteed because the degree of the polynomial equation is even. Even when the real solution set is nonempty, it contains at least two elements whose x_1 has two options.

The upper triangular equation with all nonzero diagonal entries can be solved by an analogous algorithm called back substitution [49, Chapter 3].

Algorithm 6.2 (Back Substitution). *If $\mathcal{U} \in \mathbb{C}^{n \times n \times \cdots \times n}$ is upper triangular and $\mathbf{b} \in \mathbb{C}^n$, then this algorithm overwrites $\mathbf{b}$ with one of the solutions to $\mathcal{U}\mathbf{x}^{m-1} = \mathbf{b}$.*

$b_n \leftarrow$ one of the $(m-1)$-st roots of $b_n / u_{nn\cdots n}$

for $i = n-1 : -1 : 1$

    for $k = 1 : m$

$$p_k \leftarrow \sum \left\{ u_{i i_2 \ldots i_m} \cdot \prod_{\substack{l=2,\ldots,m \\ l \neq p_1,\ldots,p_{k-1}}} b_{i_l} \; : \; i_2, \ldots, i_m \geq i, \right.$$

$$\left. i_{p_1}, \ldots, i_{p_{k-1}} \text{ are the only } k-1 \text{ indices equaling } i \right\}$$

    end

    $b_i \leftarrow$ one of the roots of $p_1 + p_2 t + \cdots + p_m t^{m-1} = b_i$

end

A higher-order triangular equation with some zero diagonal entries can also have solutions, which is different from the matrix case. As we can see in the algorithms, the polynomial equation of degree $m - 1$ reduces to a lower-degree one but still has the possibility of being feasible.

Now we consider the triangular equation associated with a nonsingular $\mathcal{M}$-tensor, that is, its diagonal entries are positive and its off-diagonal entries are nonpositive. Take the lower triangular $\mathcal{M}$-equation $\mathcal{L}\mathbf{x}^{m-1} = \mathbf{b}$ as an example. We proved in the Chapter 5 that the diagonal entries of $\mathcal{L}$ are positive; furthermore, from the definitions of lower triangular tensors and nonsingular $\mathcal{M}$-tensors, we know that the entries of $\mathcal{L}$ in the off-lower triangular part are zeros and the off-diagonal entries in the lower triangular part are nonpositive.

Proposition 6.1. *Let $\mathcal{L}$ be an mth-order n-dimensional lower triangular $\mathcal{M}$-tensor. If $\mathbf{b}$ is nonnegative, then $\mathcal{L}\mathbf{x}^{m-1} = \mathbf{b}$ has at least one nonnegative solution. Furthermore, if $\mathbf{b}$ is positive, then $\mathcal{L}\mathbf{x}^{m-1} = \mathbf{b}$ has a unique positive solution.*

Proof. When we solve the equation by Algorithm 6.1, the coefficients $p_1, p_2, \ldots, p_{m-1}$ are nonpositive and p_m is positive in each step. Thus the companion matrix [49, Page 382] of $p_1 + p_2 t + \cdots + p_m t^{m-1} = b_i$ $(i = 1, 2, \ldots, n)$:

$$C_i = \begin{bmatrix} 0 & 1 & 0 & \cdots & 0 \\ 0 & 0 & 1 & \cdots & 0 \\ \vdots & \vdots & \vdots & \ddots & \vdots \\ 0 & 0 & 0 & \cdots & 1 \\ \frac{b_i - p_1}{p_m} & \frac{-p_2}{p_m} & \frac{-p_3}{p_m} & \cdots & \frac{-p_{m-1}}{p_m} \end{bmatrix},$$

is a nonnegative matrix when $b_i \geq 0$, and it is an irreducible nonnegative matrix, which is not similar via a permutation to a block upper triangular matrix [9, Chapter 2], when $b_i > 0$. Therefore the polynomial equation in each step has at least one nonnegative solution $x_i = \rho(C_i)$ if $\mathbf{b}$ is nonnegative, and it has a unique positive solution if $\mathbf{b}$ is positive. This indicates that a triangular $\mathcal{M}$-equation with a nonnegative right-hand side has at least one nonnegative solution $\mathbf{x} \in (\mathcal{L}^{-1}\mathbf{b})_+$ and a triangular $\mathcal{M}$-equation with a positive right-hand side has a unique positive solution $\mathbf{x} \in (\mathcal{L}^{-1}\mathbf{b})_{++}$. $\square$

We call the solution $\mathbf{x}$ in the above proof the *standard solution* to $\mathcal{L}\mathbf{x}^{m-1} = \mathbf{b}$. Let $\widehat{\mathbf{b}}$ and $\widetilde{\mathbf{b}}$ be two nonnegative vectors satisfying $\mathbf{0} \leq \widehat{\mathbf{b}} \leq \widetilde{\mathbf{b}}$. Denote the standard solution to $\mathcal{L}\mathbf{x}^{m-1} = \widehat{\mathbf{b}}$ and $\mathcal{L}\mathbf{x}^{m-1} = \widetilde{\mathbf{b}}$ as $\widehat{\mathbf{x}}$ and $\widetilde{\mathbf{x}}$, respectively. It is apparent that $\widehat{x}_1 \leq \widetilde{x}_1$. Assume that $\widehat{x}_k \leq \widetilde{x}_k$ for $k = 1, 2, \ldots, i - 1$. Then the coefficients of the polynomial equation in the i-th step satisfy that $0 \geq \widehat{p}_j \geq \widetilde{p}_j$

for $j = 1, 2, \ldots, m - 1$ and $\widehat{p}_m = \widetilde{p}_m$. Then $0 \le \widehat{C}_i \le \widetilde{C}_i$ in the i-th step, thus $\widehat{x}_i = \rho(\widehat{C}_i) \le \rho(\widetilde{C}_i) = \widetilde{x}_i$ [9, Corollary 1.5]. We conclude that $\mathbf{0} \le \widehat{\mathbf{x}} \le \widetilde{\mathbf{x}}$ when $\mathbf{0} \le \widehat{\mathbf{b}} \le \widetilde{\mathbf{b}}$.

6.3 $\mathcal{M}$-Equations and Beyond

We shall investigate the multilinear equations whose coefficient tensors are $\mathcal{M}$-tensors or some related structured tensors in this section.

6.3.1 $\mathcal{M}$-Equations

An $\mathcal{M}$-equation is referred to

$$(6.4) \qquad \qquad \mathcal{A}\mathbf{x}^{m-1} = \mathbf{b},$$

where $\mathcal{A} = s\mathcal{I} - \mathcal{B}$ is an $\mathcal{M}$-tensor. Particularly, when the right-hand side $\mathbf{b}$ is nonnegative, we also require a nonnegative solution.

Consider the fixed-point iteration

$$(6.5) \qquad \mathbf{x}_{k+1} = T_{\mathcal{A}}(\mathbf{x}_k) := (s^{-1}\mathcal{B}\mathbf{x}_k^{m-1} + s^{-1}\mathbf{b})^{[1/(m-1)]}, \ \ k = 0, 1, 2, \ldots.$$

It is easy to understand that each fixed point of this iteration is a solution of (6.4), and vice versa.

To study this fixed-point iteration, we need some concepts about cones. Let $\mathbb{E}$ be a real Banach space. If $\mathbb{P}$ is a nonempty closed and convex set in $\mathbb{E}$ where

- $\mathbf{x} \in \mathbb{P}$ and $\lambda \ge 0$ imply $\lambda\mathbf{x} \in \mathbb{P}$, and

- $\mathbf{x} \in \mathbb{P}$ and $-\mathbf{x} \in \mathbb{P}$ imply $\mathbf{x} = \mathbf{0}$,

then $\mathbb{P}$ is called a *cone* in $\mathbb{E}$. Furthermore, a cone $\mathbb{P}$ induces a semi-order in $\mathbb{E}$, that is, $\mathbf{x} \le \mathbf{y}$ if $\mathbf{y} - \mathbf{x} \in \mathbb{P}$. Let $\{\mathbf{x}_n\}$ be an arbitrary increasing series in $\mathbb{E}$ with an upper bound, that is, there exists $\mathbf{y} \in \mathbb{E}$ such that

$$\mathbf{x}_1 \le \mathbf{x}_2 \le \cdots \le \mathbf{x}_n \le \cdots \le \mathbf{y}.$$

If there exists $\mathbf{x}_* \in \mathbb{E}$ such that $\|\mathbf{x}_n - \mathbf{x}_*\| \to 0$ $(n \to \infty)$, then we call $\mathbb{P}$ a *regular cone*. Let $T : \mathbb{D} \to \mathbb{E}$ be a map, where $\mathbb{D}$ is a subset in $\mathbb{E}$. If $\mathbf{x} \le \mathbf{y}$ $(\mathbf{x}, \mathbf{y} \in \mathbb{D})$ implies $T(\mathbf{x}) \le T(\mathbf{y})$, then we call T an increasing map on $\mathbb{D}$. Apparently, $T_{\mathcal{A}}$ is an *increasing map* on $\mathbb{R}_+^n$. For an increasing map on a regular cone, we have the following fixed-point theorem:

Theorem 6.2 ([2]). *Let $\mathbb{P}$ be a regular cone in an ordered Banach space $\mathbb{E}$ and $[\mathbf{u}, \mathbf{v}] \subset \mathbb{E}$ be a bounded order interval. Suppose that $T : [\mathbf{u}, \mathbf{v}] \to \mathbb{E}$ is an increasing continuous map which satisfies*

$$(6.6) \qquad \qquad \mathbf{u} \le T(\mathbf{u}), \quad \mathbf{v} \ge T(\mathbf{v}).$$

Then T has at least one fixed point in $[\mathbf{u}, \mathbf{v}]$. Moreover, there exists a minimal fixed point $\mathbf{x}_*$ and a maximal fixed point $\mathbf{x}^*$ in the sense that every fixed point $\bar{\mathbf{x}}$ satisfies $\mathbf{x}_* \le \bar{\mathbf{x}} \le \mathbf{x}^*$. Finally, the iterative method

$$(6.7) \qquad \qquad \mathbf{x}_{k+1} = T(\mathbf{x}_k), \quad k = 0, 1, 2, \ldots$$

converges to $\mathbf{x}_*$ *from below if* $\mathbf{x}_0 = \mathbf{u}$

(6.8) $$\mathbf{u} = \mathbf{x}_0 \le \mathbf{x}_1 \le \cdots \le \mathbf{x}_*,$$

and to $\mathbf{x}^*$ *from above if* $\mathbf{x}_0 = \mathbf{v}$

(6.9) $$\mathbf{v} = \mathbf{x}_0 \ge \mathbf{x}_1 \ge \cdots \ge \mathbf{x}^*.$$

By the above fixed-point theorem, we can study the existence of the positive solutions of the $\mathcal{M}$-equations. We have the following theorem, which reveals an important and interesting property of nonsingular $\mathcal{M}$-tensors.

Theorem 6.3. *If* $\mathcal{A}$ *is a nonsingular $\mathcal{M}$-tensor, then for every positive vector* $\mathbf{b}$ *the multilinear system of equations* $\mathcal{A}\mathbf{x}^{m-1} = \mathbf{b}$ *has a unique positive solution.*

Proof. Recall that a nonnegative solution of (6.4) is exactly a fixed point of

(6.10) $$T_{\mathcal{A}} : \mathbb{R}^n_+ \to \mathbb{R}^n_+, \ \mathbf{x} \mapsto \left(s^{-1}\mathcal{B}\mathbf{x}^{m-1} + s^{-1}\mathbf{b}\right)^{[1/(m-1)]}.$$

Note that $\mathbb{R}^n_+$ is a regular cone and $T_{\mathcal{A}}$ is an increasing continuous map. When $\mathcal{A}$ is a nonsingular $\mathcal{M}$-tensor, that is, $s > \rho(\mathcal{B})$, there exists a positive vector $\mathbf{z} \in \mathbb{R}^n_{++}$ such that $\mathcal{A}\mathbf{z}^{m-1} > \mathbf{0}$. Denote

$$\underline{\gamma} = \min_{i=1,2,\dots,n} \frac{b_i}{(\mathcal{A}\mathbf{z}^{m-1})_i}, \quad \overline{\gamma} = \max_{i=1,2,\dots,n} \frac{b_i}{(\mathcal{A}\mathbf{z}^{m-1})_i}.$$

Then $\underline{\gamma}\mathcal{A}\mathbf{z}^{m-1} \le \mathbf{b} \le \overline{\gamma}\mathcal{A}\mathbf{z}^{m-1}$, which indicates that

$$\underline{\gamma}^{1/(m-1)}\mathbf{z} \le T_{\mathcal{A}}\left(\underline{\gamma}^{1/(m-1)}\mathbf{z}\right), \quad \overline{\gamma}^{1/(m-1)}\mathbf{z} \ge T_{\mathcal{A}}\left(\overline{\gamma}^{1/(m-1)}\mathbf{z}\right).$$

By Theorem 6.2, there exists at least one fixed point $\bar{\mathbf{x}}$ of $T_{\mathcal{A}}$ with

$$\underline{\gamma}^{1/(m-1)}\mathbf{z} \le \bar{\mathbf{x}} \le \overline{\gamma}^{1/(m-1)}\mathbf{z},$$

which is obviously a positive vector when $\mathbf{b}$ is positive.

Furthermore, we can prove that the positive fixed point is unique when $\mathbf{b}$ is positive. Assume that there are two positive fixed points $\mathbf{x}$ and $\mathbf{y}$, that is,

$$T_{\mathcal{A}}(\mathbf{x}) = \mathbf{x} > \mathbf{0}, \quad T_{\mathcal{A}}(\mathbf{y}) = \mathbf{y} > \mathbf{0}.$$

Denote $\eta = \min_{i=1,2,\dots,n} \frac{x_i}{y_i}$, thus $\mathbf{x} \ge \eta\mathbf{y}$ and $x_j = \eta y_j$ for some j. If $\eta < 1$, then $\mathcal{A}(\eta\mathbf{y})^{m-1} = \eta^{m-1}\mathbf{b} < \mathbf{b}$, which indicates

$$T_{\mathcal{A}}(\eta\mathbf{y}) = \left[s^{-1}\mathcal{B}(\eta\mathbf{y})^{m-1} + s^{-1}\mathbf{b}\right]^{[1/(m-1)]} > \eta\mathbf{y}.$$

However, since $T_{\mathcal{A}}$ is nonnegative and increasing, we have

$$T_{\mathcal{A}}(\eta\mathbf{y})_j \le T_{\mathcal{A}}(\mathbf{x})_j = x_j = \eta y_j.$$

This is a contradiction. Thus $\eta \ge 1$, which implies $\mathbf{x} \ge \mathbf{y}$. Similarly, we can also show that $\mathbf{y} \ge \mathbf{x}$, so $\mathbf{x} = \mathbf{y}$. Therefore the positive fixed point of $T_{\mathcal{A}}$, and equivalently the positive solution to $\mathcal{A}\mathbf{x}^{m-1} = \mathbf{b}$, is unique. $\square$

Combining Theorem 3 in [38], we can rewrite the above theorem into an equivalent condition for nonsingular $\mathcal{M}$-tensors, which generalizes the nonnegative inverse property of M-matrix [9, Chapter 6] to the tensor case.

Theorem 6.4. *Let $\mathcal{A}$ be a $\mathcal{Z}$-tensor. Then it is a nonsingular $\mathcal{M}$-tensor if and only if $(\mathcal{A}^{-1}\mathbf{b})_{++}$ has a unique element for every positive vector $\mathbf{b}$.*

Proof. On the one hand, if $\mathcal{A}$ is a nonsingular $\mathcal{M}$-tensor, then by Theorem 6.3 we have the existence and uniqueness of the element in the positive solution set $(\mathcal{A}^{-1}\mathbf{b})_{++}$. On the other hand, if $(\mathcal{A}^{-1}\mathbf{b})_{++}$ has a unique element for every positive vector $\mathbf{b}$, then there must be a positive vector $\mathbf{x}$ such that $\mathcal{A}\mathbf{x}^{m-1} > 0$. Thus by [38, Theorem 3], the $\mathcal{Z}$-tensor $\mathcal{A}$ must also be a nonsingular $\mathcal{M}$-tensor. $\qquad\square$

We introduce the notation $\mathcal{A}_{++}^{-1}\mathbf{b}$ to denote the unique positive solution to $\mathcal{A}\mathbf{x}^{m-1} = \mathbf{b}$ for a nonsingular $\mathcal{M}$-tensor $\mathcal{A}$ and a positive vector $\mathbf{b}$. Then $\mathcal{A}_{++}^{-1} : \mathbb{R}_{++}^n \to \mathbb{R}_{++}^n$ is an increasing map under the partial order $\geq$ in the cone $\mathbb{R}_{++}^n$, that is, $\mathcal{A}_{++}^{-1}\widehat{\mathbf{b}} \geq \mathcal{A}_{++}^{-1}\widetilde{\mathbf{b}} > 0$ if $\widehat{\mathbf{b}} \geq \widetilde{\mathbf{b}} > 0$.

Furthermore, when $\mathcal{A} = s\mathcal{I} - \mathcal{B}$ is a singular $\mathcal{M}$-tensor, that is, $s = \rho(\mathcal{B})$, we also discuss the equation $\mathcal{A}\mathbf{x}^{m-1} = \mathbf{b}$. By a similar derivation, we have the following result about general $\mathcal{M}$-equations.

Theorem 6.5. *Let $\mathcal{A}$ be an $\mathcal{M}$-tensor. If there exists a nonnegative vector $\mathbf{v}$ such that $\mathcal{A}\mathbf{v}^{m-1} \geq \mathbf{b}$, then $(\mathcal{A}^{-1}\mathbf{b})_+$ is nonempty.*

Generally speaking, the nonnegative solution set $(\mathcal{A}^{-1}\mathbf{b})_+$ in the above theorem has more than one element, and these nonnegative solutions lay on a hypersurface in $\mathbb{R}^n$. The following example illustrates this phenomenon.

Example 6.1. We first construct a $3 \times 3 \times 3$ singular $\mathcal{M}$-tensor $\mathcal{A} = \mathcal{I} - \mathcal{B}$, where $\mathcal{B}$ is a nonnegative tensor with $\mathcal{B}\mathbf{1}^2 = \mathbf{1}$, so that $\rho(\mathcal{B}) = 1$, and $b_{ijk} = 0$ if $i \in \{2,3\}$ and either $j = 1$ or $k = 1$. Apparently, the subtensor $\mathcal{A}_{2:3,2:3,2:3}$ is also a singular $\mathcal{M}$-tensor, which satisfies that $\mathcal{A}_{2:3,2:3,2:3}\mathbf{1}^2 = \mathbf{0}$. Therefore for each right-hand side $\mathbf{b} = (b_1, 0, 0)^\top$, the nonnegative solutions to $\mathcal{A}\mathbf{x}^2 = \mathbf{b}$ have the form $\mathbf{x} = (\alpha, \beta, \beta)^\top$. We display the procedures of the iteration $\mathbf{x}_{k+1} = (\mathcal{B}\mathbf{x}_k^2 + \mathbf{b})^{[1/2]}$ with two kinds of initial points $(\beta, \beta, \beta)^\top$ and $(2 - \beta, \beta, \beta)^\top$ $(0 \leq \beta \leq 1)$ in Fig. 6.1. The square points associated with the initial points $(\beta, \beta, \beta)^\top$, the triangular points with $(2 - \beta, \beta, \beta)^\top$, and the solid points stand for some nonnegative solutions. We can see that every iteration converges to a nonnegative solution along a line, which verifies the results of Theorem 6.2. Moreover, all the solutions lie on a curve.

6.3.2   Nonpositive Right-Hand Side

We also consider the nonsingular $\mathcal{M}$-equations with nonpositive right-hand sides. When the coefficient tensor is of even order, the situation is simple. Assume that $\mathcal{A}$ is an even-order nonsingular $\mathcal{M}$-tensor and $\mathbf{b}$ is a nonpositive vector. Then the equation $\mathcal{A}\mathbf{x}^{m-1} = \mathbf{b}$ is equivalent to $\mathcal{A}(-\mathbf{x})^{m-1} = -\mathbf{b}$, which is a nonsingular $\mathcal{M}$-equation with a nonnegative right-hand side.

However, the case is totally different when the coefficient tensor is of odd order. To analyze the odd-order nonsingular $\mathcal{M}$-equations with nonpositive right-hand sides, we need the following equivalent definition of nonsingular $\mathcal{M}$-tensors, which generalizes condition (C12) in the previous chapter.

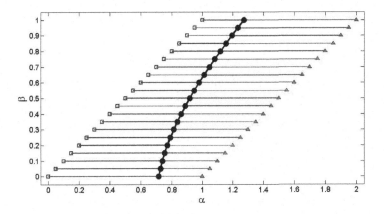

Figure 6.1: The iterations for a singular $\mathcal{M}$-equation.

Theorem 6.6. *Let $\mathcal{A}$ be a $\mathcal{Z}$-tensor. Then $\mathcal{A}$ is a nonsingular $\mathcal{M}$-tensor if and only if $\mathcal{A}$ does not reverse the sign of any vector $\mathbf{x}$; that is, if $\mathbf{x} \neq \mathbf{0}$ and $\mathbf{b} = \mathcal{A}\mathbf{x}^{m-1}$, then for some index i,*

$$x_i^{m-1}b_i > 0.$$

Proof. If $\mathcal{A}$ is a nonsingular $\mathcal{M}$-tensor, then we show that $\mathcal{A}$ does not reverse the sign of any vector $\mathbf{x}$. Assume that $\mathbf{x} \neq \mathbf{0}$ and $\mathbf{b} = \mathcal{A}\mathbf{x}^{m-1}$ with $x_i^{m-1}b_i \leq 0$ for all i. Let J be the largest index set such that $x_j \neq 0$ for all $j \in J$, and $\mathcal{A}_J$ is the corresponding leading sub-tensor of $\mathcal{A}$. Then

$$\mathbf{b}_J = \mathcal{A}_J\mathbf{x}_J^{m-1}.$$

Since $x_i^{m-1}b_i \leq 0$ and $x_j \neq 0$ for all $j \in J$, there exists a nonnegative diagonal tensor $\mathcal{D}$ such that

$$\mathbf{b}_J = -\mathcal{D}\mathbf{x}_J^{m-1}.$$

Thus we have $(\mathcal{A}_J + \mathcal{D})\mathbf{x}_J^{m-1} = 0$, which is a contradiction.

If $\mathcal{A}$ is not a nonsingular $\mathcal{M}$-tensor, then it has an eigenvector $\mathbf{x}$ such that $\mathcal{A}\mathbf{x}^{m-1} = \lambda\mathbf{x}^{[m-1]}$ and $\lambda \leq 0$. Thus $\mathcal{A}$ reverses the sign of vector $\mathbf{x}$. Therefore if $\mathcal{A}$ does not reverse the sign of any vector, then $\mathcal{A}$ must be a nonsingular $\mathcal{M}$-tensor. □

From the above theorem, we can see that there is no real vector $\mathbf{x}$ such that $\mathbf{b} = \mathcal{A}\mathbf{x}^{m-1}$ is nonpositive when m is odd, since $\mathbf{x}^{[m-1]}$ is always nonnegative.

6.3.3 Nonhomogeneous Left-Hand Side

All the multilinear equations that we have discussed have homogeneous left-hand sides. However, similar results can be obtained for some special equations with nonhomogeneous left-hand sides. Consider the following equation

$$\mathcal{A}\mathbf{x}^{m-1} - \mathcal{B}_{m-1}\mathbf{x}^{m-2} - \cdots - \mathcal{B}_2\mathbf{x} = \mathbf{b} > 0,$$

where $\mathcal{A} = s\mathcal{I} - \mathcal{B}_m$ is an mth-order nonsingular $\mathcal{M}$-tensor and $\mathcal{B}_p$ is a pth-order nonnegative tensor for $p = 2, 3, \ldots, m$. We further assume that there exists a positive vector $\mathbf{v}$ such that

$$(6.11) \qquad \mathcal{A}\mathbf{v}^{m-1} - \mathcal{B}_{m-1}\mathbf{v}^{m-2} - \cdots - \mathcal{B}_2\mathbf{v} > \mathbf{0},$$

which is an extension of a parallel property of nonsingular $\mathcal{M}$-tensors. Similarly, we discuss the fixed-point iteration

$$\mathbf{x}_k = F(\mathbf{x}_{k-1}), \ \ k = 1, 2, \ldots$$

for analyzing the positive solution, where

$$F(\mathbf{x}) = \left[s^{-1}(\mathcal{B}_m\mathbf{x}^{m-1} + \mathcal{B}_{m-1}\mathbf{x}^{m-2} + \cdots + \mathcal{B}_2\mathbf{x} + \mathbf{b}) \right]^{[1/(m-1)]}.$$

Thus a total analogous discussion can be conducted to this fixed-point iteration, which leads to the conclusion that this nonhomogeneous equation has a unique positive solution for each positive right-hand side. Actually, condition (6.11) can be further simplified, and we have the following theorem about this kind of equations with nonhomogeneous left-hand sides.

Theorem 6.7. *Let $\mathcal{A}$ be an mth-order $\mathcal{Z}$-tensor and $\mathcal{B}_p$ be a pth-order nonnegative tensor for $p = 2, 3, \ldots, m - 1$. Then the equation*

$$(6.12) \qquad \mathcal{A}\mathbf{x}^{m-1} - \mathcal{B}_{m-1}\mathbf{x}^{m-2} - \cdots - \mathcal{B}_2\mathbf{x} = \mathbf{b}$$

has a unique positive solution for every positive vector $\mathbf{b}$ if and only if $\mathcal{A}$ is a nonsingular $\mathcal{M}$-tensor.

Proof. If the equation (6.12) has a positive solution $\mathbf{x}$ for a positive $\mathbf{b}$, then $\mathcal{A}\mathbf{x}^{m-1} = \mathcal{B}_{m-1}\mathbf{x}^{m-2} + \cdots + \mathcal{B}_2\mathbf{x} + \mathbf{b} \geq \mathbf{b} > \mathbf{0}$ since $\mathcal{B}_p$ $(p = 2, 3, \ldots, m - 1)$ are nonnegative tensors. Hence $\mathcal{A}$ is a nonsingular $\mathcal{M}$-tensor.

   If $\mathcal{A}$ is a nonsingular $\mathcal{M}$-tensor, then there exists a positive vector $\mathbf{v}$ such that $\mathcal{A}\mathbf{v}^{m-1} > \mathbf{0}$. Since the order of $\mathcal{A}$ is higher than the orders of those nonnegative tensors $\mathcal{B}_p$, there exists a scalar α such that $\mathcal{A}(\alpha\mathbf{v})^{m-1} > \mathcal{B}_{m-1}(\alpha\mathbf{v})^{m-2} + \cdots + \mathcal{B}_2(\alpha\mathbf{v})$, which indicates that the condition (6.11) is satisfied. Thus we know from the above discussion that the equation (6.12) has a unique positive solution for every positive vector $\mathbf{b}$. $\qquad \square$

6.3.4   Absolute $\mathcal{M}$-Equations

The above results for $\mathcal{M}$-tensors can also be extended. A tensor is called an *absolute $\mathcal{M}$-tensor*, if it can be written as $\mathcal{A} = s\mathcal{I} - \mathcal{B}$ with $s > \rho(|\mathcal{B}|)$, that is, $s\mathcal{I} - |\mathcal{B}|$ is a nonsingular $\mathcal{M}$-tensor. It can be verified directly that an absolute $\mathcal{M}$-tensor must be a nonsingular $\mathcal{H}$-tensor, since $s - |b_{ii\ldots i}| \leq |s - b_{ii\ldots i}|$. An *absolute $\mathcal{M}$-equation* is referred to

$$\mathcal{A}\mathbf{x}^{m-1} = \mathbf{b},$$

where $\mathcal{A}$ is an absolute $\mathcal{M}$-tensor. To deal with this situation, we employ the Brouwer fixed-point theorem [105] instead of Theorem 6.2.

Theorem 6.8 (Brouwer). *Let Ω be a bounded closed convex set in $\mathbb{R}^n$. If map $F : \Omega \to \Omega$ is continuous, then there exists $\mathbf{x}_* \in \Omega$ such that $F(\mathbf{x}_*) = \mathbf{x}_*$.*

When m is even, we also consider the fixed-point iteration

$$\mathbf{x}_k = F(\mathbf{x}_{k-1}) := (s^{-1}\mathcal{B}\mathbf{x}_{k-1}^{m-1} + s^{-1}\mathbf{b})^{[1/(m-1)]}, \ k = 1, 2, \ldots.$$

Applying Brouwer's fixed point theorem, we can prove that $F(\mathbf{x})$ must have a fixed point in $\mathbb{R}^n$. Since $s\mathcal{I} - |\mathcal{B}|$ is a nonsingular $\mathcal{M}$-tensor, there exists a vector $\mathbf{v} > \mathbf{0}$ such that $(s\mathcal{I} - |\mathcal{B}|)\mathbf{v}^{m-1} > \mathbf{0}$, that is, $s^{-1}|\mathcal{B}|\mathbf{v}^{m-1} < \mathbf{v}^{[m-1]}$. Let $\alpha \in (0, 1)$ satisfy that $s^{-1}|\mathcal{B}|\mathbf{v}^{m-1} \le \alpha\mathbf{v}^{[m-1]}$. Take

$$\Omega = \{\mathbf{x} \in \mathbb{R}^n : |\mathbf{x}| \le \gamma\mathbf{v}\},$$

where γ satisfies

$$\alpha(\gamma\mathbf{v})^{[m-1]} + s^{-1}|\mathbf{b}| \le (\gamma\mathbf{v})^{[m-1]}.$$

Note that there must exist such a γ, since $\alpha < 1$ and $s^{-1}|\mathbf{b}|$ is a fixed vector. Then for every $\mathbf{x} \in \Omega$,

$$|F(\mathbf{x})| \le (s^{-1}|\mathcal{B}||\mathbf{x}|^{m-1} + s^{-1}|\mathbf{b}|)^{[1/(m-1)]} \le [\alpha(\gamma\mathbf{v})^{[m-1]} + s^{-1}|\mathbf{b}|]^{[1/(m-1)]} \le \gamma\mathbf{v}.$$

Therefore $F(\cdot)$ is a continuous map from Ω to Ω. Thus there exists $\mathbf{x}_* \in \Omega$ such that $F(\mathbf{x}_*) = \mathbf{x}_*$, that is, $\mathcal{A}\mathbf{x}_*^{m-1} = \mathbf{b}$. It is equivalent to say that an even order absolute $\mathcal{M}$-equation has at least one real solution for every real right-hand side.

6.3.5 Banded $\mathcal{M}$-Equation

The inverse matrix of a banded nonsingular M-matrix has a *decay property* [8, 16], that is, its entries decay exponentially from the diagonal to the corners. It is interesting that a banded nonsingular $\mathcal{M}$-tensor has a similar property. Although there is no inverse tensor for a general $\mathcal{M}$-tensor, we can express this property in terms of the minimal nonnegative solution to $\mathcal{A}\mathbf{x}^{m-1} = \mathbf{e}_p$, decaying exponentially from the p-th entry to both ends.

Without loss of generality, let $\mathcal{A} = \mathcal{I} - \mathcal{B}$ be an mth-order nonsingular $\mathcal{M}$-tensor, which is strictly diagonally dominant, that is, $\mathcal{A}\mathbf{1}^2 > \mathbf{0}$. Furthermore, we also assume that $\mathcal{B}$ is d-banded, that is, $b_{i_1 i_2 \ldots i_m} = 0$ if $(i_1, i_2, \ldots, i_m)$ lies out of

$$\Omega_{i_1} := \{(i_1, i_2, \ldots, i_m) : |i_s - i_t| \le d \text{ for all } s, t = 1, 2, \ldots, m\}.$$

We shall explore the structure of the minimal nonnegative solution to $\mathcal{A}\mathbf{x}^{m-1} = \mathbf{e}_p$. Consider the iteration

$$\mathbf{x}_{k+1} = (\mathcal{B}\mathbf{x}_k^{m-1} + \mathbf{e}_p)^{[1/(m-1)]}, \ k = 0, 1, 2, \ldots$$

with the initial vector $\mathbf{x}_0 = \mathbf{0}$, which converges to the minimal nonnegative solution $\mathbf{x}_*$ by Theorem 6.2. We know that $\{\mathbf{x}_k\}_{k=0}^{\infty}$ is increasing, that is, $\mathbf{0} = \mathbf{x}_0 \le \mathbf{x}_1 \le \mathbf{x}_2 \le \cdots \le \mathbf{x}_*$.

Denote the function corresponding to the i-th row of the tensor $\mathcal{B}$:

$$f_i(\xi) = \sum_{(i, i_2, \ldots, i_m) \in \Omega_i} b_{i i_2 \ldots i_m} \xi^{|i_2 - p| + \cdots + |i_m - p|} - \xi^{(m-1)|i-p|}.$$

Notice that $f_i(1) < 0$ since $\mathcal{A}$ is strictly diagonally dominant, and $f_i(\xi) > 0$ when $\xi > 0$ is sufficiently closed to 0. Thus there exists a root of $f_i(\xi) = 0$ in the interval $(0, 1)$, then we denote

$$\widehat{\xi} = \max\{\xi \in (0, 1) : f_i(\xi) = 0 \text{ for an arbitrary } i = 1, 2, \ldots, n\}.$$

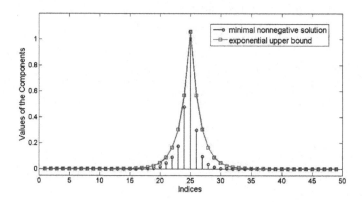

Figure 6.2: The decay property of the minimal nonnegative solution to $\mathcal{A}\mathbf{x}^2 = \mathbf{e}_{25}$.

First, it is apparent that the initial vector $\mathbf{x}_0 = \mathbf{0}$ satisfies $(\mathbf{x}_0)_i \leq (\mathbf{x}_0)_p \cdot \widehat{\xi}^{|i-p|}$ for all i. Next, assume that $\mathbf{x}_k$ satisfies that $(\mathbf{x}_k)_i \leq (\mathbf{x}_k)_p \cdot \widehat{\xi}^{|i-p|}$ for all i. Then we have for the $(k+1)$-th step

$$
\begin{aligned}
(\mathbf{x}_{k+1})_i &= \sum_{(i,i_2,\ldots,i_m)\in\Omega_i} b_{ii_2\ldots i_m}(\mathbf{x}_k)_{i_2}\cdots(\mathbf{x}_k)_{i_m} \\
&\leq \sum_{(i,i_2,\ldots,i_m)\in\Omega_i} b_{ii_2\ldots i_m}\widehat{\xi}^{|i_2-p|+\cdots+|i_m-p|}(\mathbf{x}_k)_p^{m-1} \\
&\leq \left(\sum_{(i,i_2,\ldots,i_m)\in\Omega_i} b_{ii_2\ldots i_m}\widehat{\xi}^{|i_2-p|+\cdots+|i_m-p|}\right)(\mathbf{x}_{k+1})_p^{m-1} \\
&\leq \widehat{\xi}^{(m-1)|i-p|}\cdot(\mathbf{x}_{k+1})_p^{m-1}.
\end{aligned}
$$

Therefore the fixed point $\mathbf{x}_*$ also satisfies $(\mathbf{x}_*)_i \leq (\mathbf{x}_*)_p \cdot \widehat{\xi}^{|i-p|}$ for all i, which implies the decay property. We shall employ an example to illustrate this interesting property.

Example 6.2. We first construct a $49\times49\times49$ nonsingular $\mathcal{M}$-tensor which is 1-banded. Then we compute both the minimal nonnegative solution $\mathbf{x}_*$ to $\mathcal{A}\mathbf{x}^2 = \mathbf{e}_{25}$ and the exponential upper bound $(\mathbf{x}_*)_{25}\cdot\widehat{\xi}^{|i-25|}$ for $i = 1,2,\ldots,49$. From Fig. 6.2, we can see that the solution is perfectly controlled by the exponential upper bound, which illustrates the trend of decay very well.

6.4   Iterative Methods for $\mathcal{M}$-Equations

Consider the nonsingular $\mathcal{M}$-equation $\mathcal{A}\mathbf{x}^{m-1} = \mathbf{b} > \mathbf{0}$, where $\mathcal{A} = s\mathcal{I} - \mathcal{B}$ is a nonsingular $\mathcal{M}$-tensor, that is, $\mathcal{B}$ is nonnegative and $s > \rho(\mathcal{B})$. What we need is a positive solution to this equation. Since there is no "LU factorization" for general tensors, we propose several iterative methods for this multilinear equation.

Iterative Methods	Alternatives of $\mathcal{M}$
Jacobi	diagonal part
forward G-S	lower triangular part
simplified forward G-S	strictly lower triangular part & diagonal part
backward G-S	upper triangular part
simplified backward G-S	strictly upper triangular part & diagonal part

Table 6.1: Iterative methods for nonsingular $\mathcal{M}$-equations.

6.4.1 The Classical Iterations

Recalling the Jacobi and the Gauss-Seidel methods for linear equations, we split the coefficient matrix and apply a fixed-point iteration to obtain the solution. Similarly, we split the coefficient tensor $\mathcal{A} = \mathcal{M} - \mathcal{N}$ so that the equations with coefficient tensor $\mathcal{M}$ are easy to solve and $\mathcal{N}$ is nonnegative. Then the iteration

$$\mathbf{x}_k = \mathcal{M}_{++}^{-1}(\mathcal{N}\mathbf{x}_{k-1}^{m-1} + \mathbf{b}), \ k = 1, 2, \ldots$$

offers a nonnegative solution if it converges, which is equivalent to $\mathcal{M}\mathbf{x}_k^{m-1} = \mathcal{N}\mathbf{x}_{k-1}^{m-1} + \mathbf{b}$. Under the requirement that the multilinear equation $\mathcal{M}\mathbf{y}^{m-1} = \mathbf{g}$ is easy to solve, we have five alternatives as in Table 6.1.

We now study the convergence of this iterative scheme. Consider an operator $\phi : \mathbb{R}^n \to \mathbb{R}^n$. Let $\mathbf{x}_*$ be a fixed point of $\phi(\mathbf{x})$. Then we call $\mathbf{x}_*$ an *attracting fixed point*, if there exists $\delta > 0$ such that the sequence $\{\mathbf{x}_k\}$ defined by $\mathbf{x}_{k+1} = \phi(\mathbf{x}_k)$ converges to $\mathbf{x}_*$ for any $\mathbf{x}_0$ such that $\|\mathbf{x}_0 - \mathbf{x}_*\| \leq \delta$.

Theorem 6.9 ([105]). *Let $\mathbf{x}_*$ be a fixed point of $\phi : \mathbb{R}^n \to \mathbb{R}^n$, and let $\nabla\phi : \mathbb{R}^n \to \mathbb{R}^{n \times n}$ be the Jacobian of ϕ. Then $\mathbf{x}_*$ is an attracting fixed point if $\sigma := \rho(\nabla\phi(\mathbf{x}_*)) < 1$; further, if $\sigma > 0$, then the convergence of $\mathbf{x}_{k+1} = \phi(\mathbf{x}_k)$ to $\mathbf{x}_*$ is linear with rate σ.*

We first derive the Jacobian of the operator

$$\phi(\mathbf{x}) = \mathcal{M}_{++}^{-1}(\mathcal{N}\mathbf{x}^{m-1} + \mathbf{b}).$$

Notice that $\mathcal{A}$ can always be modified into another tensor $\widetilde{\mathcal{A}}$ such that $\mathcal{A}\mathbf{x}^{m-1} = \widetilde{\mathcal{A}}\mathbf{x}^{m-1}$ for all $\mathbf{x} \in \mathbb{R}^n$ and $\widetilde{\mathcal{A}}$ is symmetric on the last $(m-1)$ modes. Hence it is reasonable to assume that $\mathcal{A}$ is symmetric on the last $(m-1)$ modes, so that the gradient $\nabla(\mathcal{A}\mathbf{x}^{m-1}) = (m-1)\mathcal{A}\mathbf{x}^{m-2}$. Differentiating both sides of

$$\mathcal{M}\phi(\mathbf{x})^{m-1} = \mathcal{N}\mathbf{x}^{m-1} + \mathbf{b},$$

we obtain

$$\mathcal{M}\phi(\mathbf{x})^{m-2} \cdot \nabla\phi(\mathbf{x}) = \mathcal{N}\mathbf{x}^{m-2}.$$

When we substitute $\mathbf{x} = \mathbf{x}_*$, then the matrix $\mathcal{M}\phi(\mathbf{x}_*)^{m-2} = \mathcal{M}\mathbf{x}_*^{m-2}$ is a nonsingular M-matrix, since $\mathbf{x}_* > \mathbf{0}$ and

$$\mathcal{M}\mathbf{x}_*^{m-2} \cdot \mathbf{x}_* = \mathcal{N}\mathbf{x}_*^{m-1} + \mathbf{b} \geq \mathbf{b} > \mathbf{0}.$$

Therefore the Jacobian of $\phi(\mathbf{x})$ at $\mathbf{x}_*$ is

$$\nabla\phi(\mathbf{x}_*) = (\mathcal{M}\mathbf{x}_*^{m-2})^{-1}\mathcal{N}\mathbf{x}_*^{m-2},$$

which is nonnegative. Since

$$\mathcal{N}\mathbf{x}_*^{m-2} \cdot \mathbf{x}_* = \mathcal{M}\mathbf{x}_*^{m-1} - \mathbf{b} \le \theta \mathcal{M}\mathbf{x}_*^{m-1}$$

with $0 \le \theta < 1$, then $\nabla\phi(\mathbf{x}_*) \cdot \mathbf{x}_* \le \theta^{1/(m-1)}\mathbf{x}_*$. Therefore the spectral radius $\rho(\nabla\phi(\mathbf{x}_*)) \le \theta^{1/(m-1)} < 1$, which indicates that $\mathbf{x}_*$ is an attracting fixed point of ϕ.

The above discussion shows that the solution is attainable. However an initial vector that can ensure convergence is required for the algorithm. Since $\mathcal{A}$ is a nonsingular $\mathcal{M}$-tensor, we can take an initial vector $\mathbf{x}_0$ satisfying

$$0 < \mathcal{A}\mathbf{x}_0^{m-1} \le \mathbf{b},$$

then we shall prove that the iteration

$$\mathbf{x}_k = \mathcal{M}_{++}^{-1}(\mathcal{N}\mathbf{x}_{k-1}^{m-1} + \mathbf{b}), \ k = 1, 2, \dots$$

converges to the solution to the positive nonsingular $\mathcal{M}$-equation with a positive right-hand side.

First, $\mathcal{A}\mathbf{x}_0^{m-1} > \mathbf{0}$ indicates $\mathcal{N}\mathbf{x}_0^{m-1} < \mathcal{M}\mathbf{x}_0^{m-1}$. Thus there is $\alpha \in (0,1)$ such that $\mathcal{N}\mathbf{x}_0^{m-1} \le \alpha(\mathcal{M}\mathbf{x}_0^{m-1})$. Moreover, let the positive number β satisfy that $\mathbf{b} \le \beta(\mathcal{M}\mathbf{x}_0^{m-1})$. The first iteration step implies that

$$\mathcal{M}\mathbf{x}_0^{m-1} \le \mathcal{N}\mathbf{x}_0^{m-1} + \mathbf{b} \le (\alpha + \beta)\mathcal{M}\mathbf{x}_0^{m-1},$$

which further implies that

$$\mathbf{x}_0^{[m-1]} \le \mathbf{x}_1^{[m-1]} \le (\alpha + \beta)\mathbf{x}_0^{[m-1]}$$

since $\mathcal{M}$ is a triangular $\mathcal{M}$-tensor. Next, we assume that

$$\mathbf{x}_{k-1}^{[m-1]} \le \mathbf{x}_k^{[m-1]} \le (\alpha^k + \alpha^{k-1}\beta + \cdots + \beta)\mathbf{x}_0^{[m-1]}.$$

Then the $(k+1)$-st iteration step implies that

$$\mathcal{N}\mathbf{x}_{k-1}^{m-1} + \mathbf{b} \le \mathcal{N}\mathbf{x}_k^{m-1} + \mathbf{b} \le [\alpha(\alpha^k + \alpha^{k-1}\beta + \cdots + \beta) + \beta]\mathcal{M}\mathbf{x}_0^{m-1},$$

thus

$$\mathbf{x}_k^{[m-1]} \le \mathbf{x}_{k+1}^{[m-1]} \le (\alpha^{k+1} + \alpha^k\beta + \cdots + \beta)\mathbf{x}_0^{[m-1]}.$$

Therefore we can conclude that the sequence $\{\mathbf{x}_k\}$ is increasing and with an upper bound $\left(\frac{\beta}{1+\alpha}\right)^{1/(m-1)}\mathbf{x}_0$. It converges to a positive vector $\mathbf{x}_*$, which is the positive solution to $\mathcal{A}\mathbf{x}^{m-1} = \mathbf{b}$.

An SOR-like acceleration [49] can also be applied to this iterative method. For instance, we can choose a proper $\omega > 0$ so that the iterative scheme

$$\mathbf{x}_k = (\mathcal{M} - \omega\mathcal{I})_{++}^{-1}[(\mathcal{N} - \omega\mathcal{I})\mathbf{x}_{k-1}^{m-1} + \mathbf{b}], \ k = 1, 2, \dots$$

converges faster than the original scheme. The acceleration effect is due to a smaller α in the above discussion. There are some restrictions when choosing the parameter ω, such as

1. $\omega > 0$,

2. $\mathcal{M} - \omega\mathcal{I}$ is still a nonsingular $\mathcal{M}$-tensor,

3. $(\mathcal{N} - \omega\mathcal{I})\mathbf{x}_{k-1}^{m-1} + \mathbf{b} > \mathbf{0}$ for all $k = 1, 2, \dots$.

However, whether there is an optimal parameter ω and its estimation remains open questions.

6.4.2 The Newton Method for Symmetric $\mathcal{M}$-Equations

When $\mathcal{A}$ is a symmetric nonsingular $\mathcal{M}$-tensor, we can have other methods to compute the positive solution to $\mathcal{A}\mathbf{x}^{m-1} = \mathbf{b} > \mathbf{0}$. It is proven in [133] that a symmetric nonsingular $\mathcal{M}$-tensor is strictly copositive [101], that is, $\mathcal{A}\mathbf{x}^m > 0$ for all $\mathbf{x} > \mathbf{0}$. Consider the functional

$$\varphi(\mathbf{x}) := \frac{1}{m}\mathcal{A}\mathbf{x}^m - \mathbf{x}^\top \mathbf{b}.$$

Since $\varphi(\mathbf{x})$ is convex on $\Omega := \{\mathbf{x} \in \mathbb{R}^n_{++} : \mathcal{A}\mathbf{x}^{m-1} > 0\}$ and its gradient is

$$\nabla\varphi(\mathbf{x}) = \mathcal{A}\mathbf{x}^{m-1} - \mathbf{b} =: -\mathbf{r},$$

computing the positive solution to the above symmetric $\mathcal{M}$-equation is equivalent to solving the optimization problem

$$\min_{\mathbf{x}\in\Omega} \varphi(\mathbf{x}).$$

We employ the Newton method. Notice that the second-order gradient of $\varphi(\mathbf{x})$ is

$$\nabla^2\varphi(\mathbf{x}) = (m-1)\mathcal{A}\mathbf{x}^{m-2}.$$

Note that when $\mathcal{A}\mathbf{x}^{m-1} > 0$, matrix $\mathcal{A}\mathbf{x}^{m-2}$ is obviously a symmetric Z-matrix and

$$\mathcal{A}\mathbf{x}^{m-2} \cdot \mathbf{x} = \mathcal{A}\mathbf{x}^{m-1} > 0.$$

Therefore $\mathcal{A}\mathbf{x}^{m-2}$ is a symmetric nonsingular M-matrix, and thus a positive definite matrix [9, Chapter 2]. Then the Newton step

$$\mathbf{p}_k = -\left[\nabla^2\varphi(\mathbf{x}_k)\right]^{-1}\nabla\varphi(\mathbf{x}_k) = \tfrac{1}{m-1}(\mathcal{A}\mathbf{x}_k^{m-2})^{-1}\mathbf{r}_k$$

is descending. Then the iterations are as follows:

$$\begin{cases} M_k = \mathcal{A}\mathbf{x}_k^{m-2}, \\ \mathbf{r}_k = \mathbf{b} - M_k\mathbf{x}_k, \\ \mathbf{p}_k = \frac{1}{m-1}M_k^{-1}\mathbf{r}_k, \\ \mathbf{x}_{k+1} = \mathbf{x}_k + \lambda_k\mathbf{p}_k, \end{cases} \quad k = 0, 1, 2, \ldots,$$

where λ_k is chosen to ensure $\mathbf{x}_{k+1} \in \Omega$.

The Newton method is generally expensive. However, the computational cost is relatively cheap for higher-order tensor situations. The first reason is that there is no need of any extra computations for forming the matrix $M_k = \mathcal{A}\mathbf{x}_k^{m-2}$, since it is an intermediate product of the computation of $\mathcal{A}\mathbf{x}_k^{m-1}$. The next reason is that the computational complexity of solving a linear system is $\mathcal{O}(n^3)$, which is no larger than $\mathcal{O}(n^m)$, the computational complexity of computing a tensor-vector product $\mathcal{A}\mathbf{x}^{m-1}$, when $m \geq 3$.

It should be pointed out that the restriction $\mathbf{x} > \mathbf{0}$ in the optimization problem is automatically satisfied in the procedure by setting the initial vector $\mathbf{x}_0$ such that

$$\mathbf{x}_0 > \mathbf{0}, \quad \mathcal{A}\mathbf{x}_0^{m-1} > \mathbf{0},$$

which can be proven by rewriting

$$\mathbf{x}_{k+1} = M_k^{-1}\left(\tfrac{m-2}{m-1}\mathcal{A}\mathbf{x}_k^{m-1} + \tfrac{1}{m-1}\mathbf{b}\right) > \mathbf{0},$$

since M_k is a nonsingular M-matrix and $\mathcal{A}\mathbf{x}_k^{m-1}$ and $\mathbf{b}$ are both positive vectors. Furthermore, the above equality can be regarded as a fixed-point iteration as well, so the Newton method sometimes works for non-symmetric M-equations. However there is no theoretic guarantee for the non-symmetric cases, because the matrix $\mathcal{A}\mathbf{x}^{m-2}$ is not necessarily positive definite.

6.4.3   Numerical Tests

Example 6.3. We construct a 3rd-order nonsingular M-tensor $\mathcal{A} = s\mathcal{I} - \mathcal{B}$ as follows. First, we generate a nonnegative tensor $\mathcal{B} \in \mathbb{R}_+^{n \times n \times n}$ containing random values drawn from the standard uniform distribution on $(0, 1)$. Next, set the scalar

$$s = (1 + \varepsilon) \cdot \max_{i=1,2,\ldots,n} (\mathcal{B}\mathbf{1}^2)_i, \ \varepsilon > 0,$$

where $\mathbf{1} = (1, 1, \ldots, 1)^\top$. Obviously, $\mathcal{A}$ is a diagonally dominant $\mathcal{Z}$-tensor, that is, $\mathcal{A}\mathbf{1}^2 > \mathbf{0}$. Thus $\mathcal{A}$ is a nonsingular M-tensor. In this example, we take $n = 10$ and $\varepsilon = 0.01$.

We select the acceleration parameter:

$$\omega = \tau \cdot \min_{i=1,2,\ldots,n} a_{ii\ldots i}, \ 0 < \tau < 1.$$

This ensures the first two conditions discussed in the previous subsection and the last condition when $\mathbf{x}_{k-1}$ is close to the solution. In this experiment, we take $\tau = 0.35$, which is chosen by experience.

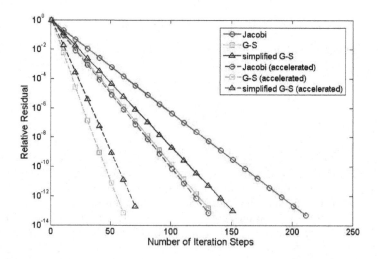

Figure 6.3: The comparison of different classical iterative methods for M-equations.

Fig. 6.3 displays the performances of six iterative algorithms for this M-equation. The convergence of this kind of iterative schemes is linear, which confirms the result of Theorem 6.9. With or without acceleration, the Gauss-Seidel method requires fewer iteration steps than the Jacobi and the simplified Gauss-Seidel methods. However, there is no need of solving polynomial equations in the Jacobi or the simplified Gauss-Seidel methods. Hence the computation cost of each step in the Gauss-Seidel method is the most expensive among these three methods. Finally, it can also be observed that the SOR-like acceleration with $\tau = 0.35$ can save more than half of the iteration steps in this experiment.

Example 6.4. We construct a symmetric $\mathcal{M}$-tensor of size $10 \times 10 \times 10$ in the following way. Let $\mathcal{B} \in \mathbb{R}^{10 \times 10 \times 10}$ be a nonnegative tensor with

$$b_{i_1 i_2 i_3} = |\tan(i_1 + i_2 + i_3)|.$$

It can be computed that $\rho(\mathcal{B}) \approx 1450.3$. Thus $\mathcal{A} = 1500\mathcal{I} - \mathcal{B}$ is a symmetric nonsingular $\mathcal{M}$-tensor. We apply the Newton, the accelerated Jacobi, the accelerated Gauss-Seidel, and the accelerated simplified Gauss-Seidel methods to this equation. The acceleration parameters for the Jacobi and Gauss-Seidel methods are taken as $\tau = 400$, which are experimentally optimal. Fig. 6.4 shows the comparison on the number of iteration steps, from which we can see that the Newton method converges much faster than other algorithms.

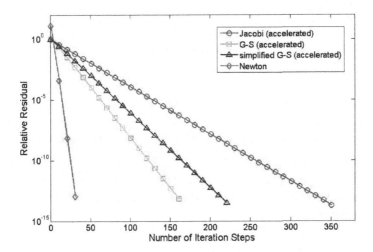

Figure 6.4: Comparison of the Newton method and other iterative methods.

Example 6.5. We take the one-dimensional case of the boundary value problem introduced in Section 1 as an example. Consider the ordinary differential equation

$$\frac{\mathrm{d}^2 x(t)}{\mathrm{d}t^2} = -\frac{f(t)}{x(t)^2} \text{ in } (0, 1),$$

with Dirichlet's boundary conditions

$$x(0) = c_0, \quad x(1) = c_1.$$

Assume that $f(t) > 0$ on $[0,1]$, $c_0, c_1 > 0$, and we require a positive solution $x(t)$ on $[0,1]$. This equation describes a particle's movement under gravitation[2]:

$$m\frac{d^2x}{dt^2} = -G\frac{Mm}{x^2} \rightarrow x^2 \cdot \frac{d^2x}{dt^2} = -GM,$$

where $G \approx 6.67 \times 10^{-11}\mathrm{N}\cdot\mathrm{m}^2/\mathrm{kg}^2$ is the gravitational constant and $M \approx 5.98 \times 10^{24}\mathrm{kg}$ is the mass of the earth.

After discretization, we get a system of polynomial equations

$$\begin{cases} x_1^3 = c_0^3, \\ 2x_i^3 - x_i^2 x_{i-1} - x_i^2 x_{i+1} = \frac{GM}{(n-1)^2}, \ i = 2,3,\ldots,n-1, \\ x_n^3 = c_1^3. \end{cases}$$

Then this can be rewritten into a multilinear equation $\mathcal{A}\mathbf{x}^3 = \mathbf{b}$, where the coefficient tensor $\mathcal{A}$ is a tensor with seven diagonals, that is, $\mathcal{A}$ is 1-banded as defined in Section 6.3.5, and

$$\begin{cases} a_{1,1,1,1} = a_{n,n,n,n} = 1, \\ a_{i,i,i,i} = 2, \ i = 2,3,\ldots,n-1, \\ a_{i,i-1,i,i} = a_{i,i,i-1,i} = a_{i,i,i,i-1} = -1/3, \ i = 2,3,\ldots,n-1, \\ a_{i,i+1,i,i} = a_{i,i,i+1,i} = a_{i,i,i,i+1} = -1/3, \ i = 2,3,\ldots,n-1, \end{cases}$$

and the right-hand side is a positive vector with

$$\begin{cases} b_1 = c_0^2, \\ b_i = \frac{GM}{(n-1)^2}, \ i = 2,3,\ldots,n-1, \\ b_n = c_1^2. \end{cases}$$

It is easy to verify that the coefficient tensor $\mathcal{A}$ is a nonsingular $\mathcal{M}$-tensor. Therefore we can solve the equation by our algorithm.

Fig. 6.5 displays the trajectory of a particle after being thrown upward near the earth surface. Assume that the distance between the surface and the earth's center is 6370 kilometers. It is well known that the trajectory can be approximated by a parabola

$$\begin{cases} x(t) = -\frac{1}{2}gt^2 + \alpha t + \beta, \\ x(0) = c_0, \ x(1) = c_1, \end{cases}$$

where $g \approx 9.8\mathrm{m/s}^2$. We conclude from Fig. 6.5 that the solution from the tensor formulation coincides with the parabolic model.

6.5 Perturbation Analysis of $\mathcal{M}$-Equations

Since we propose algorithms to solve different kinds of $\mathcal{M}$-equations, the effects of perturbations should be analyzed. Both backward errors and condition numbers are considered in this section.

---

[2]http://en.wikipedia.org/wiki/Law_of_universal_gravitation

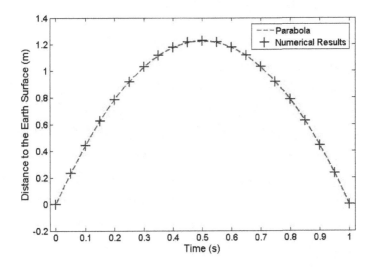

Figure 6.5: A particle's movement under the earth's gravitation.

6.5.1 Backward Errors of Triangular $\mathcal{M}$-Equations

The backward errors of triangular $\mathcal{M}$-equations are first analyzed, since triangular $\mathcal{M}$-equations can be solved by direct methods. We represent the roundoff error [49, Chapter 2] δ in

$$\mathrm{fl}(a \odot b) = (a \odot b)(1 + \delta), \quad |\delta| \leq \epsilon,$$

where $\odot$ is one of the four binary operations $+$, $-$, $*$, and $/$, and ϵ is the roundoff unit. Suppose that we use a backward stable algorithm for solving the polynomial equation

$$c_1 + c_2 t + \cdots + c_m t^{m-1} = 0,$$

so that the computed solution $\hat{t}$ can be regarded as a solution to a perturbed polynomial equation

$$c_1(1 + \gamma_1) + c_2(1 + \gamma_2)\hat{t} + \cdots + c_m(1 + \gamma_m)\hat{t}^{m-1} = 0,$$

and the errors $|\gamma_k|$ are bounded by $\gamma\epsilon$.

Now we present the backward error analysis of solving the lower triangular $\mathcal{M}$-equation $\mathcal{L}\mathbf{x}^{m-1} = \mathbf{b}$ by forward substitution. Assume that the computed positive solution $\hat{\mathbf{x}}$ is the solution to a perturbed equation

$$(\mathcal{L} + \Delta\mathcal{L})\hat{\mathbf{x}}^{m-1} = \mathbf{b}.$$

We need to bound the perturbation $\Delta\mathcal{L}$. Consider the i-th step in Algorithm 6.1. According to the above assumption, $\hat{x}_i$ satisfies that

$$(p_1 - b_i)(1 + \gamma_1) + p_2(1 + \gamma_2)\hat{x}_i + \cdots + p_m(1 + \gamma_m)\hat{x}_i^{m-1} = 0,$$

where $|\gamma_k| \leq \gamma\epsilon$ for $k = 1, 2, \ldots, m$.

Then we rearrange the above equation to

$$p_1 + p_2 \frac{1+\gamma_2}{1+\gamma_1} \cdot \widehat{x}_i + \cdots + p_m \frac{1+\gamma_m}{1+\gamma_1} \cdot \widehat{x}_i^{m-1} = b_i.$$

Moreover, from the algorithm we have

$$p_k = \mathrm{fl}\Big(\sum \{ l_{ii_2 \ldots i_m} b_{i_2} \ldots \widehat{b}_{i_{p_1}} \ldots \widehat{b}_{i_{p_{k-1}}} \ldots b_{i_m} \; : \; i_2, \ldots, i_m \leq i,$$

$$i_{p_1}, \ldots, i_{p_{k-1}} \text{ are the only } k-1 \text{ indices equaling } i \} \Big)$$

$$= \sum l_{ii_2 \ldots i_m} (1+\delta_*)^{m-k} (1+\delta_+)^{n_{ii_2 \ldots i_m}} b_{i_2} \ldots \widehat{b}_{i_{p_1}} \ldots \widehat{b}_{i_{p_{k-1}}} \ldots b_{i_m},$$

where $n_{ii_2 \ldots i_m}$ denotes the number of additions, which is bounded by the number of terms in this summation:

$$\binom{m-1}{k-1} (i-1)^{m-k}.$$

It can be verified that

$$(m-k) + \binom{m-1}{k-1}(i-1)^{m-k} \leq m-1+(n-1)^{m-1}.$$

Therefore the backward perturbation can be bounded by

$$|\Delta \mathcal{L}| \leq \Big[\big(m-1+(n-1)^{m-1}+2\gamma \big)\epsilon + o(\epsilon^2) \Big] \cdot |\mathcal{L}|,$$

from which it can be concluded that the direct method for triangular $\mathcal{M}$-equations is backward stable.

6.5.2   Condition Numbers

Before studying the condition numbers of $\mathcal{M}$-equations, we first introduce some tensor norms. The p-norm of tensor $\mathcal{A}$ is defined by

$$\|\mathcal{A}\|_p = \|\mathcal{A}_{(1)}\|_p, \quad p = 1, 2, \ldots, \infty.$$

We know that the tensor-vector product can be expressed as

$$\mathcal{A}\mathbf{x}^{m-1} = \mathcal{A}_{(1)} \cdot \widetilde{\mathbf{x}},$$

where $\widetilde{\mathbf{x}} = \underbrace{\mathbf{x} \otimes \mathbf{x} \otimes \cdots \otimes \mathbf{x}}_{m-1}$. It is direct to verify that $\|\widetilde{\mathbf{x}}\|_p = \|\mathbf{x}\|_p^{m-1}$ for $p = 1, 2, \ldots, \infty$. Thus

$$\|\mathcal{A}\mathbf{x}^m\|_p = \|\mathcal{A}_{(1)} \cdot \widetilde{\mathbf{x}}\|_p \leq \|\mathcal{A}_{(1)}\|_p \cdot \|\widetilde{\mathbf{x}}\|_p = \|\mathcal{A}_{(1)}\|_p \cdot \|\mathbf{x}\|_p^{m-1}.$$

holds. The norms used below are p-norms unless otherwise stated.

Suppose a nonsingular $\mathcal{M}$-equation $\mathcal{A}\mathbf{x}^{m-1} = \mathbf{b}$ has structured perturbations on $\mathcal{A}$ and $\mathbf{b}$ such that

$$\frac{\|\Delta \mathcal{A}\|}{\|\mathcal{A}\|} \leq \epsilon, \quad \frac{\|\Delta \mathbf{b}\|}{\|\mathbf{b}\|} \leq \epsilon,$$

$A + \Delta A$ remains a nonsingular M-tensor, and $\mathbf{b} + \Delta \mathbf{b} > \mathbf{0}$. Thus there exists $\Delta \mathbf{x}$ such that $\mathbf{x} + \Delta \mathbf{x} > \mathbf{0}$ satisfies

$$(A + \Delta A)(\mathbf{x} + \Delta \mathbf{x})^{m-1} = \mathbf{b} + \Delta \mathbf{b}.$$

Our goal is to bound the norm of $\Delta \mathbf{x}$. Expanding the left-hand side of the above equation and neglecting the high-order items of the perturbations, we get

$$(A + \Delta A)(\mathbf{x} + \Delta \mathbf{x})^{m-1} \approx (A + \Delta A)\mathbf{x}^{m-1} + (m-1)A\mathbf{x}^{m-2}\Delta \mathbf{x}.$$

Notice that $A\mathbf{x}^{m-1} = \mathbf{b}$, and we have

$$\Delta \mathbf{x} \approx \frac{1}{m-1}(A\mathbf{x}^{m-2})^{-1}(\Delta \mathbf{b} - \Delta A\mathbf{x}^{m-1}).$$

Taking norms and using the triangle inequality for vector norms and the definition of tensor norms, we get

$$\frac{\|\Delta \mathbf{x}\|}{\|\mathbf{x}\|} \lesssim \frac{\|(A\mathbf{x}^{m-2})^{-1}\|}{(m-1)\|\mathbf{x}\|}\left(\|\Delta \mathbf{b}\| + \|\Delta A\| \cdot \|\mathbf{x}\|^{m-1}\right)$$

$$= \frac{\|(A\mathbf{x}^{m-2})^{-1}\| \cdot \|\mathbf{b}\|}{(m-1)\|\mathbf{x}\|} \cdot \frac{\|\Delta \mathbf{b}\|}{\|\mathbf{b}\|} + \frac{\|(A\mathbf{x}^{m-2})^{-1}\| \cdot \|A\|}{(m-1)\|\mathbf{x}\|^{2-m}} \cdot \frac{\|\Delta A\|}{\|A\|}.$$

If we denote the *condition number*

$$\kappa(A; \mathbf{x}) := \frac{\|(A\mathbf{x}^{m-2})^{-1}\| \cdot \|A\|}{(m-1)\|\mathbf{x}\|^{2-m}},$$

and the *effective condition number* (refer to [75] for the matrix case)

$$\kappa_{\text{eff}}(A, \mathbf{b}; \mathbf{x}) := \frac{\|(A\mathbf{x}^{m-2})^{-1}\| \cdot \|\mathbf{b}\|}{(m-1)\|\mathbf{x}\|},$$

then we finally obtain the perturbation bound

$$\frac{\|\Delta \mathbf{x}\|}{\|\mathbf{x}\|} \lesssim \kappa(A; \mathbf{x}) \cdot \frac{\|\Delta A\|}{\|A\|} + \kappa_{\text{eff}}(A, \mathbf{b}; \mathbf{x}) \cdot \frac{\|\Delta \mathbf{b}\|}{\|\mathbf{b}\|}.$$

Example 6.6. We construct a 3rd-order nonsingular M-equation $A\mathbf{x}^2 = \mathbf{b}$ as in Example 6.3. Then we first assume that there is no perturbation on the right-hand side $\mathbf{b}$, and we study the relative errors on the solutions when the coefficient tensors are with structured perturbations. In fact, we randomly construct ΔA such that

$$\|\Delta A\|_\infty < \varepsilon \cdot \|A\|_\infty,$$

which ensures that $(A + \Delta A)\mathbf{1}^2 > \mathbf{0}$, and thus $A + \Delta A$ is still a nonsingular M-tensor. The dots in Fig. 6.6 display the relationship between $\|\Delta A\|_\infty / \|A\|_\infty$ and $\|\Delta \mathbf{x}\|_\infty / \|\mathbf{x}\|_\infty$, and the line in Fig. 6.6 possess a slope $\kappa(A; \mathbf{x})$, from which we can conclude that when $\Delta \mathbf{b} = \mathbf{0}$,

$$\frac{\|\Delta \mathbf{x}\|}{\|\mathbf{x}\|} \leq \kappa(A; \mathbf{x}) \cdot \frac{\|\Delta A\|}{\|A\|}.$$

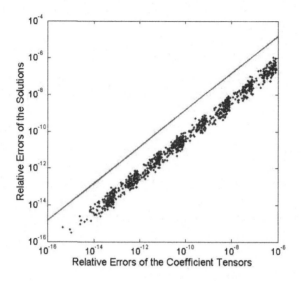

Figure 6.6: Condition Number

Similarly, we assume that there is no perturbation on the coefficient tensor $\mathcal{A}$, and we study the relative error on the solution when the right-hand side is perturbed such that

$$\mathbf{b} + \Delta\mathbf{b} > \mathbf{0}.$$

The dots in Fig. 6.7 display the relationship between $\|\Delta\mathbf{b}\|_\infty/\|\mathbf{b}\|_\infty$ and $\|\Delta\mathbf{x}\|_\infty/\|\mathbf{x}\|_\infty$, and the line in Fig. 6.7 has a slope $\kappa_{\text{eff}}(\mathcal{A}, \mathbf{b}; \mathbf{x})$, from which we can conclude that when $\Delta\mathcal{A} = \mathcal{O}$,

$$\frac{\|\Delta\mathbf{x}\|}{\|\mathbf{x}\|} \leq \kappa_{\text{eff}}(\mathcal{A}, \mathbf{b}; \mathbf{x}) \cdot \frac{\|\Delta\mathbf{b}\|}{\|\mathbf{b}\|}.$$

More generally, assume that both the coefficient tensor and the right-hand side are perturbed. We visualize the relationship among $\|\Delta\mathcal{A}\|_\infty/\|\mathcal{A}\|_\infty$, $\|\Delta\mathbf{b}\|_\infty/\|\mathbf{b}\|_\infty$ and $\|\Delta\mathbf{x}\|_\infty/\|\mathbf{x}\|_\infty$ in Fig. 6.8 by dots, and the surface in this figure is

$$z = \kappa(\mathcal{A}; \mathbf{x}) \cdot x + \kappa_{\text{eff}}(\mathcal{A}, \mathbf{b}; \mathbf{x}) \cdot y.$$

We can see that all the dots are below the surface. Therefore Fig. 6.8 verifies the perturbation bound we have derived.

6.6   Inverse Iteration

The power method for matrix eigenproblems is extended to nonnegative tensor eigenproblems, often called the NQZ method in [83], and its convergence is widely studied in [21, 131, 132, 135]. Moreover, we also have inverse iteration, or called Noda iteration for nonnegative matrices, which converges much faster than the power method in common [46, 63, 88]. We generalize the inverse

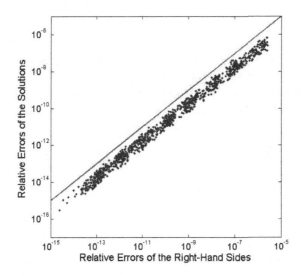

Figure 6.7: Effective Condition Number

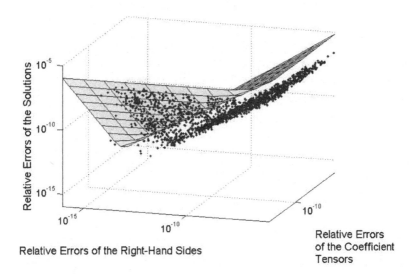

Figure 6.8: Condition Number and Effective Condition Number

iteration for solving the spectral radii of nonnegative tensors, which is based on nonsingular $\mathcal{M}$-equations.

We present the algorithm first and then give some explanations and discuss its convergence.

Algorithm 6.3 (Inverse Iteration). *If $\mathcal{B}$ is an mth-order n-dimensional non-negative tensor, then this algorithm gives the spectral radius of $\mathcal{B}$ and the corresponding eigenvector when converges.*

> $\mathbf{x}_0 =$ an approximated eigenvector
> $s_0 = (1 + \varepsilon) \cdot \max\limits_{i} (\mathcal{B}\mathbf{x}_0^{m-1})_i / (\mathbf{x}_0^{[m-1]})_i$
> for $k = 1, 2, \ldots$
> > $\mathbf{y}_k = (s_{k-1}\mathcal{I} - \mathcal{B})_{++}^{-1} \left(\mathbf{x}_{k-1}^{[m-1]} \right)$
> > if $\angle \left(\mathbf{y}_k^{[m-1]}, s_{k-1}\mathbf{y}_k^{[m-1]} - \mathbf{x}_{k-1}^{[m-1]} \right)$ is small enough
> > > break
> > end
> > $s_k = (1 + \varepsilon) \cdot \left[s_{k-1} - \left(\min\limits_{i}(\mathbf{x}_{k-1})_i / (\mathbf{y}_k)_i \right)^{m-1} \right]$
> > $\mathbf{x}_k = \mathbf{y}_k / \|\mathbf{y}_k\|$
> end

There are some remarks on this algorithm. First, the shift in the k-th step is actually [129, 128]

$$s_k = (1 + \varepsilon) \cdot \max_i \frac{(\mathcal{B}\mathbf{x}_k^{m-1})_i}{(\mathbf{x}_k^{[m-1]})_i} \geq (1 + \varepsilon) \cdot \rho(\mathcal{B}),$$

which can be easily obtained from the relationship $(s_{k-1}\mathcal{I} - \mathcal{B})\mathbf{y}_k^{m-1} = \mathbf{x}_{k-1}^{[m-1]}$. The update of s_k in the algorithm avoids an extra tensor-vector product. Next, for the same reason, the stopping rule tests the angle between $\mathbf{y}_k^{[m-1]}$ and $\mathcal{B}\mathbf{y}_k^{m-1}$ in fact. Finally, the parameter $\varepsilon \geq 0$ in the shifts is aimed at reducing the iteration steps for solving the nonsingular $\mathcal{M}$-equation in each step if we apply an iterative method. When we choose the Newton method to solve the multilinear equations, we can simply set $\varepsilon = 0$ for its fast convergence.

Now we show the convergence of this algorithm. Denote

$$\underline{\lambda}(\mathbf{x}) = \min_{1 \leq i \leq n} \frac{(\mathcal{B}\mathbf{x}^{m-1})_i}{(\mathbf{x}^{[m-1]})_i} \quad \text{and} \quad \overline{\lambda}(\mathbf{x}) = \max_{1 \leq i \leq n} \frac{(\mathcal{B}\mathbf{x}^{m-1})_i}{(\mathbf{x}^{[m-1]})_i}$$

for a positive vector $\mathbf{x}$. Then $\underline{\lambda}(\mathbf{x}) \leq \rho(\mathcal{B}) \leq \overline{\lambda}(\mathbf{x})$, since $\mathcal{B}$ is nonnegative [129, 128]. From the iteration, we have

$$\mathbf{x}_{k-1}^{[m-1]} = (s_{k-1}\mathcal{I} - \mathcal{B})\mathbf{y}_k^{m-1} \leq (s_{k-1} - \underline{\lambda}(\mathbf{y}_k))\mathbf{y}_k^{[m-1]},$$

and thus

$$\underline{\lambda}(\mathbf{x}_{k-1})(s_{k-1}\mathcal{I} - \mathcal{B})\mathbf{y}_k^{m-1} = \underline{\lambda}(\mathbf{x}_{k-1})\mathbf{x}_{k-1}^{[m-1]} \leq \mathcal{B}\mathbf{x}_{k-1}^{m-1} \leq (s_{k-1} - \underline{\lambda}(\mathbf{y}_k))\mathcal{B}\mathbf{y}_k^{m-1}.$$

Rearrange the above inequality, and we can write

$$\mathcal{B}\mathbf{y}_k^{m-1} \geq \frac{\underline{\lambda}(\mathbf{x}_{k-1})s_{k-1}}{s_{k-1} - \underline{\lambda}(\mathbf{y}_k) + \underline{\lambda}(\mathbf{x}_{k-1})}\mathbf{y}_k^{m-1}.$$

	I		II		III		IV	
	Steps	Time	Steps	Time	Steps	Time	Steps	Time
Inverse	4	7.38	6	11.00	5	9.89	5	0.58
Power	11	2.65	33	7.93	41	9.90	NC	NC
Shifted	–	–	–	–	–	–	25	1.41

Table 6.2: Compare of the inverse iteration and the (shifted) power method.

From the definition of $\underline{\lambda}(\mathbf{y}_k)$, it holds that

$$\frac{\underline{\lambda}(\mathbf{x}_{k-1})s_{k-1}}{s_{k-1} - \underline{\lambda}(\mathbf{y}_k) + \underline{\lambda}(\mathbf{x}_{k-1})} \leq \underline{\lambda}(\mathbf{y}_k),$$

which indicates that $\underline{\lambda}(\mathbf{x}_k) = \underline{\lambda}(\mathbf{y}_k) \geq \underline{\lambda}(\mathbf{x}_{k-1})$. Analogously, we can verify that $\overline{\lambda}(\mathbf{x}_k) \leq \overline{\lambda}(\mathbf{x}_{k-1})$. Therefore the nested sequence of intervals $[\underline{\lambda}(\mathbf{x}_k), \overline{\lambda}(\mathbf{x}_k)]$ converges to $[\underline{\lambda}^*, \overline{\lambda}^*]$. It is apparent that $\rho(\mathcal{B}) \in [\underline{\lambda}^*, \overline{\lambda}^*]$. Thus when $\underline{\lambda}^* = \overline{\lambda}^* = \rho(\mathcal{B})$, $\mathbf{x}_k$ converges to the corresponding eigenvector.

Example 6.7. We employ the following four tensors to compare the performance of the inverse iteration with other existing algorithms for nonnegative tensor eigenproblems.

I. As in Example 6.3, we generate a nonnegative tensor $\mathcal{B} \in \mathbb{R}_+^{10 \times 10 \times 10}$ containing random values drawn from the standard uniform distribution on $(0, 1)$. This is an irreducible example.

II. We also generate a nonnegative tensor $\mathcal{B} \in \mathbb{R}_+^{10 \times 10 \times 10}$ containing random values drawn from the standard uniform distribution on $(0, 1)$ first. Then set $b_{i_1 i_2 i_3} = 0$ if $i_1 \geq 7$ and $i_2, i_3 < 7$. Thus this tensor is reducible.

III. The third test tensor is the same as the one in Example 6.4, that is, a nonnegative tensor $\mathcal{B} \in \mathbb{R}_+^{10 \times 10 \times 10}$ with $b_{i_1 i_2 i_3} = |\tan(i_1 + i_2 + i_3)|$, which is symmetric.

IV. The last example was given by Liping Zhang in one of her talks [3] to show that the power method, also called the NQZ method, does not always converge for nonnegative tensors. Let $\mathcal{B} \in \mathbb{R}_+^{3 \times 3 \times 3}$ with $b_{133} = b_{233} = b_{311} = b_{322} = 1$ and other entries being zeros. Since the power method does not work for this example, we compare the inverse iteration with the shifted power method [67, 135], and we select an experimentally optimal shift. Moreover, the multilinear equation $(s\mathcal{I} - \mathcal{B})\mathbf{x}^2 = \mathbf{b}$ is equivalent to a linear equation

$$\begin{bmatrix} s & 0 & -1 \\ 0 & s & -1 \\ -1 & -1 & s \end{bmatrix} \cdot \begin{bmatrix} x_1^2 \\ x_2^2 \\ x_3^2 \end{bmatrix} = \begin{bmatrix} b_1 \\ b_2 \\ b_3 \end{bmatrix},$$

which can be solved by the LU factorization.

---

[3] http://www.nim.nankai.edu.cn/activites/conferences/hy20120530/index.htm

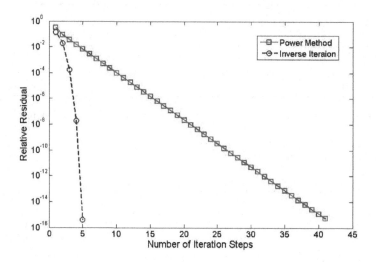

Figure 6.9: The comparison of inverse iteration and power method.

Since the shifted power method has no apparent advantages over the original power method for the first three examples, we just compare the inverse iteration with the power method. From Table 6.2, we conclude that the inverse iteration always converges faster than the (shifted) power method. However, since there is a multilinear equation to solve in each step, the inverse iteration will take up more time when the power method also has a relatively fast convergence. But when the power method converges slowly (as in the third example) or the multilinear equation can be easily solved (as in the fourth example), the inverse iteration performs better than the (shifted) power method.

Fig. 6.9 displays the convergence of the inverse iteration and the power method for the third example. From this figure, we can expect that the inverse iteration converges quadratically, as for the matrix case. Nevertheless, we cannot prove this conjecture and it remains an open question.

Appendix

Consider the following the ordinary differential equation with Dirichlets boundary condition

$$\begin{cases} u(x)^{m-2} \cdot u''(x) = -f(x), \ x \in (0,1), \\ u(0) = g_0, \ u(1) = g_1, \end{cases}$$

where $f(x) > 0$ in $(0,1)$ and $g_0, g_1 > 0$.

Partition the interval $[0,1]$ into $n-1$ small intervals with the same length

$h = 1/(n-1)$, and denote

$$\mathbf{u}_h = \Big[u(0), u(h), \ldots, u\big((n-1)h\big)\Big]^{\mathsf{T}},$$

$$\mathbf{v}_h = \Big[-u^{(4)}(0), -u^{(4)}(h), \ldots, -u^{(4)}\big((n-1)h\big)\Big]^{\mathsf{T}},$$

$$\mathbf{f}_h = \Big[g_0/h^2, f(h), \ldots, f\big((n-2)h\big), g_1/h^2\Big]^{\mathsf{T}},$$

where $u(x)$ is the exact solution of the above boundary-value problem.

The discretization tensor $\mathcal{L}_h$ of the operator $u \mapsto u^{m-2} \cdot u''$ is introduced in the first section, and our numerical solution $\widehat{\mathbf{u}}_h$ is obtained by solving the unique positive solution of an $\mathcal{M}$-equation $\mathcal{L}_h \widehat{\mathbf{u}}_h^{m-1} = \mathbf{f}_h$. It is well known that the truncation error of the discretization is

$$\tfrac{u(x-h)-2u(x)+u(x+h)}{h^2} - u''(x) = \tfrac{h^2}{12} u^{(4)}(x) + O(h^4).$$

Thus we have

$$\mathcal{L}_h \mathbf{u}_h^{m-1} - \mathcal{L}_h \widehat{\mathbf{u}}_h^{m-1} = \mathcal{L}_h \mathbf{u}_h^{m-1} - \mathbf{f}_h = \tfrac{h^2}{12} \cdot \mathbf{u}_h^{[m-2]} \circ \mathbf{v}_h + O(h^4),$$

which further implies that

$$d_h(\mathbf{u}_h, \widehat{\mathbf{u}}_h) := \big\|\mathcal{L}_h \mathbf{u}_h^{m-1} - \mathcal{L}_h \widehat{\mathbf{u}}_h^{m-1}\big\|_\infty \le \tfrac{h^2}{12} \cdot \|u\|_{L^\infty}^{m-2} \cdot \|u^{(4)}\|_{L^\infty} + O(h^4).$$

It can be verified that $d_h(\cdot, \cdot)$ is a metric in the cone $\{\mathbf{x} > \mathbf{0} : \mathcal{L}_h \mathbf{x}^{m-1} > \mathbf{0}\}$. Then we can say that the numerical solution $\widehat{\mathbf{u}}_h$ is very close to the exact solution $\mathbf{u}_h$ when the parameter h is small enough. Next, we shall estimate the convergence of the discretization scheme.

Note that $\mathcal{L}_h \mathbf{u}_h^{m-1}$ is a positive vector when h is small enough, then the matrix $\mathcal{L}_h \mathbf{u}_h^{m-2}$ is a nonsingular M-matrix as discussed in Section 4. Hence we have the first-order approximation

$$\mathbf{u}_h - \widehat{\mathbf{u}}_h \approx (\mathcal{L}_h \mathbf{u}_h^{m-2})^{-1}(\mathcal{L}_h \mathbf{u}_h^{m-1} - \mathcal{L}_h \widehat{\mathbf{u}}_h^{m-1})$$

when h is small enough, and thus

$$\|\mathbf{u}_h - \widehat{\mathbf{u}}_h\|_\infty \lesssim \big\|(\mathcal{L}_h \mathbf{u}_h^{m-2})^{-1}\big\|_\infty \cdot \big\|\mathcal{L}_h \mathbf{u}_h^{m-1} - \mathcal{L}_h \widehat{\mathbf{u}}_h^{m-1}\big\|_\infty.$$

We thus need to bound the ∞-norm of the inverse of the M-matrix $\mathcal{L}_h \mathbf{u}_h^{m-2}$. First denote $\mathcal{L}_h = s_h \mathcal{I} - \mathcal{A}_h$, where $s_h = 2/h^2$ and $\mathcal{A}_h$ is nonnegative. Then we can write

$$\begin{aligned}
(\mathcal{L}_h \mathbf{u}_h^{m-2})^{-1} &= \big((s_h \mathcal{I} - \mathcal{A}_h) \mathbf{u}_h^{m-2}\big)^{-1} \\
&= \big(s_h U_h^{m-2} - \mathcal{A}_h \mathbf{u}_h^{m-2}\big)^{-1} \\
&= s_h^{-1} U_h \Big[I - s_h^{-1} U_h^{-(m-1)} (\mathcal{A}_h \mathbf{u}_h^{m-2}) U_h\Big]^{-1} U_h^{-(m-1)},
\end{aligned}$$

where $U_h = \mathrm{diag}\big((\mathbf{u}_h)_1, (\mathbf{u}_h)_2, \ldots, (\mathbf{u}_h)_n\big)$.

Denote $W_h = U_h^{-(m-1)}(\mathcal{A}_h \mathbf{u}_h^{m-2})U_h$, which is a nonnegative matrix. Note that $(\mathcal{A}_h \mathbf{u}_h^{m-2})U_h \mathbf{1} = \mathcal{A}_h \mathbf{u}_h^{m-1}$, thus the summation of all the rows of W_h:

$$
\begin{aligned}
W_h \mathbf{1} &= U_h^{-(m-1)} \mathcal{A}_h \mathbf{u}_h^{m-1} \le \mathbf{1} \cdot \max_{i=1:n} \frac{(\mathcal{A}_h \mathbf{u}_h^{m-1})_i}{(\mathbf{u}_h)_i^{m-1}} \\
&= \mathbf{1} \cdot \max_{i=1:n} \frac{s_h(\mathbf{u}_h)_i^{m-1} - (\mathcal{L}_h \mathbf{u}_h^{m-1})_i}{(\mathbf{u}_h)_i^{m-1}} \\
&= \mathbf{1} \cdot \left[s_h - \min_{i=1:n} \frac{(\mathcal{L}_h \mathbf{u}_h^{m-1})_i}{(\mathbf{u}_h)_i^{m-1}} \right] \\
&=: \mathbf{1} \cdot (s_h - \gamma_h).
\end{aligned}
$$

Similarly, we have $W_h^k \mathbf{1} \le W_h^{k-1} \mathbf{1} \cdot (s_h - \gamma_h) \le \cdots \le \mathbf{1} \cdot (s_h - \gamma_h)^k$. Also, because $W_h \mathbf{1} \le \mathbf{1} \cdot (s_h - \gamma_h) < \mathbf{1} \cdot s_h$ and W_h is an irreducible nonnegative matrix, we have $\rho(s_h^{-1} W_h) < 1$. Applying the Taylor expansion of the matrix $(I - X)^{-1} = I + X + X^2 + \dots$ for $\rho(X) < 1$, we obtain

$$
\left(I - s_h^{-1} W_h\right)^{-1} \mathbf{1} = \sum_{k=0}^{\infty} \left(s_h^{-1} W_h\right)^k \mathbf{1} \le \sum_{k=0}^{\infty} \mathbf{1} \cdot (1 - \gamma_h/s_h)^k = \mathbf{1} \cdot (s_h/\gamma_h).
$$

Finally, we get an upper bound of the ∞-norm of the nonnegative matrix $(\mathcal{L}_h \mathbf{u}_h^{m-2})^{-1}$:

$$
\begin{aligned}
\|(\mathcal{L}_h \mathbf{u}_h^{m-2})^{-1}\|_\infty &= \max_{i=1:n} \left((\mathcal{L}_h \mathbf{u}_h^{m-2})^{-1} \mathbf{1}\right)_i \\
&= \max_{i=1:n} \left(s_h^{-1} U_h \left(I - s_h^{-1} W_h\right)^{-1} U_h^{-(m-1)} \mathbf{1}\right)_i \\
&\le \left(\min_{i=1:n} \mathbf{u}_h\right)^{-(m-1)} \cdot \max_{i=1:n} \frac{(\mathbf{u}_h)_i^{m-1}}{(\mathcal{L}_h \mathbf{u}_h^{m-1})_i} \cdot \max_{i=1:n} \mathbf{u}_h \\
&\approx \left(\min_{i=1:n} \mathbf{u}_h\right)^{-(m-1)} \cdot \max_{i=1:n} \frac{(\mathbf{u}_h)_i^{m-1}}{(\mathbf{f}_h)_i} \cdot \max_{i=1:n} \mathbf{u}_h \\
&\le \frac{\max_x u(x)^m}{\min_x u(x)^{m-1} \cdot \min_x f(x)}.
\end{aligned}
$$

Note that $u(x)$ can also be regarded as the solution of the elliptic problem

$$
\begin{cases}
u''(x) = f_1(x) := -f(x)/u(x)^{m-2}, & x \in (0,1), \\
u(0) = g_0, \ u(1) = g_1.
\end{cases}
$$

Therefore we know that $u(x) \ge \min\{g_0, g_1\}$ since $f(x) > 0$ and $g_0, g_1 > 0$. Then

$$
\|\mathbf{u}_h - \hat{\mathbf{u}}_h\|_\infty \lesssim \frac{\|u\|_\infty^m}{\min\{g_0, g_1\}^{m-1} \cdot \min_x f(x)} \cdot \frac{h^2}{12} \cdot \|u\|_{L^\infty}^{m-2} \cdot \|u^{(4)}\|_{L^\infty} =: Kh^2,
$$

where the constant K is independent with the parameter h.

Therefore the sequence $\{\hat{\mathbf{u}}_h\}$ converges to the exact solution when $h \to 0$.

Bibliography

[1] N. I. Akhiezer. *The Classical Moment Problem and Some Related Questions in Analysis.* Translated by N. Kemmer. Hafner Publishing Co., New York, 1965.

[2] H. Amann. Fixed point equations and nonlinear eigenvalue problems in ordered Banach spaces. *SIAM Rev.*, 18(4):620–709, 1976.

[3] P. Aubry, D. Lazard, and M. Moreno Maza. On the theories of triangular sets. *J. Symbolic Comput.*, 28(1-2):105–124, 1999.

[4] P. Aubry and M. Moreno Maza. Triangular sets for solving polynomial systems: a comparative implementation of four methods. *J. Symbolic Comput.*, 28(1-2):125–154, 1999.

[5] R. Badeau and R. Boyer. Fast multilinear singular value decomposition for structured tensors. *SIAM J. Matrix Anal. Appl.*, 30(3):1008–1021, 2008.

[6] Z. Bai, J. Demmel, J. Dongarra, A. Ruhe, and H. van der Vorst, editors. *Templates for the Solution of Algebraic Eigenvalue Problems: A Practical Guide*, volume 11 of *Software, Environments, and Tools*. Society for Industrial and Applied Mathematics (SIAM), Philadelphia, PA, 2000.

[7] F. L. Bauer and C. T. Fike. Norms and exclusion theorems. *Numer. Math.*, 2:137–141, 1960.

[8] M. Benzi and G. H. Golub. Bounds for the entries of matrix functions with applications to preconditioning. *BIT*, 39(3):417–438, 1999.

[9] A. Berman and R. J. Plemmons. *Nonnegative Matrices in the Mathematical Sciences*, volume 9 of *Classics in Applied Mathematics*. Society for Industrial and Applied Mathematics (SIAM), Philadelphia, PA, 1994. Revised reprint of the 1979 original.

[10] D. A. Bini, G. Latouche, and B. Meini. *Numerical Methods for Structured Markov Chains.* Numerical Mathematics and Scientific Computation. Oxford University Press, New York, 2005.

[11] M. Boizard, R. Boyer, G. Favier, and P. Larzabal. Fast multilinear singular value decomposition for higher-order hankel tensors. In *Sensor Array and Multichannel Signal Processing Workshop (SAM), 2014 IEEE 8th*, pages 437–440. IEEE, 2014.

Theory and Computation of Tensors.
http://dx.doi.org/10.1016/B978-0-12-803953-3.50010-1

[12] D. L. Boley, F. T. Luk, and D. Vandevoorde. Vandermonde factorization of a Hankel matrix. In *Scientific computing (Hong Kong, 1997)*, pages 27–39. Springer, Singapore, 1997.

[13] R. Boyer, L. De Lathauwer, and K. Abed-Meraim. Higher order tensor-based method for delayed exponential fitting. *IEEE Transactions on Signal Processing*, 55(6):2795–2809, 2007.

[14] K. Browne, S. Qiao, and Y. Wei. A Lanczos bidiagonalization algorithm for Hankel matrices. *Linear Algebra Appl.*, 430(5-6):1531–1543, 2009.

[15] C. Bu, X. Zhang, J. Zhou, W. Wang, and Y. Wei. The inverse, rank and product of tensors. *Linear Algebra Appl.*, 446:269–280, 2014.

[16] C. Canuto, V. Simoncini, and M. Verani. On the decay of the inverse of matrices that are sum of Kronecker products. *Linear Algebra Appl.*, 452:21–39, 2014.

[17] R. H.-F. Chan and X.-Q. Jin. *An Introduction to Iterative Toeplitz Solvers*, volume 5 of *Fundamentals of Algorithms*. Society for Industrial and Applied Mathematics (SIAM), Philadelphia, PA, 2007.

[18] K. Chang, L. Qi, and T. Zhang. A survey on the spectral theory of nonnegative tensors. *Numer. Linear Algebra Appl.*, 20(6):891–912, 2013.

[19] K. C. Chang, K. Pearson, and T. Zhang. Perron-Frobenius theorem for nonnegative tensors. *Commun. Math. Sci.*, 6(2):507–520, 2008.

[20] K. C. Chang, K. Pearson, and T. Zhang. On eigenvalue problems of real symmetric tensors. *J. Math. Anal. Appl.*, 350(1):416–422, 2009.

[21] K.-C. Chang, K. J. Pearson, and T. Zhang. Primitivity, the convergence of the NQZ method, and the largest eigenvalue for nonnegative tensors. *SIAM J. Matrix Anal. Appl.*, 32(3):806–819, 2011.

[22] C. Chen and M. Moreno Maza. Algorithms for computing triangular decomposition of polynomial systems. *J. Symbolic Comput.*, 47(6):610–642, 2012.

[23] H. Chen, G. Li, and L. Qi. Further results on Cauchy tensors and Hankel tensors. *Appl. Math. Comput.*, 275:50–62, 2016.

[24] H. Chen, G. Li, and L. Qi. SOS tensor decomposition: Theory and applications. *Commun. Math. Sci.*, 2016. to appear.

[25] L. Chen, L. Han, and L. Zhou. Computing tensor eigenvalues via homotopy methods. *SIAM J. Matrix Anal. Appl.*, 37:290–319, 2016.

[26] Y. Chen, L. Qi, and Q. Wang. Computing eigenvalues of large scale hankel tensors. *J. Sci. Comput.*, 2016. doi: 10.1007/s10915-015-0155-8.

[27] Z. Chen and L. Qi. Circulant tensors with applications to spectral hypergraph theory and stochastic process. *J. Ind. Manag. Optim.*, 12:1227–1247, 2016.

[28] W.-K. Ching, X. Huang, M. K. Ng, and T.-K. Siu. *Markov Chains: Models, Algorithms and Applications.* International Series in Operations Research & Management Science, 189. Springer, New York, second edition, 2013.

[29] J. Cooper and A. Dutle. Spectra of uniform hypergraphs. *Linear Algebra Appl.*, 436(9):3268–3292, 2012.

[30] C.-F. Cui, Y.-H. Dai, and J. Nie. All real eigenvalues of symmetric tensors. *SIAM J. Matrix Anal. Appl.*, 35(4):1582–1601, 2014.

[31] C. D'Andrea and A. Dickenstein. Explicit formulas for the multivariate resultant. *J. Pure Appl. Algebra*, 164(1-2):59–86, 2001.

[32] P. J. Davis. *Circulant Matrices.* John Wiley & Sons, New York-Chichester-Brisbane, 1979. A Wiley-Interscience Publication, Pure and Applied Mathematics.

[33] L. De Lathauwer. *Signal Processing Based on Multilinear Algebra.* PhD thesis, Katholike Universiteit Leuven, 1997.

[34] L. De Lathauwer. Blind separation of exponential polynomials and the decomposition of a tensor in rank-$(L_r, L_r, 1)$ terms. *SIAM J. Matrix Anal. Appl.*, 32(4):1451–1474, 2011.

[35] L. De Lathauwer, B. De Moor, and J. Vandewalle. A multilinear singular value decomposition. *SIAM J. Matrix Anal. Appl.*, 21(4):1253–1278, 2000.

[36] L. De Lathauwer, B. De Moor, and J. Vandewalle. On the best rank-1 and rank-$(R_1, R_2, \cdots, R_N)$ approximation of higher-order tensors. *SIAM J. Matrix Anal. Appl.*, 21(4):1324–1342, 2000.

[37] L. De Lathauwer and L. Hoegaerts. Rayleigh quotient iteration for the computation of the best rank-$(R_1, R_2, \ldots, R_N)$ approximation in multilinear algebra. Technical report, SCD-SISTA 04-003.

[38] W. Ding, L. Qi, and Y. Wei. $\mathcal{M}$-tensors and nonsingular $\mathcal{M}$-tensors. *Linear Algebra Appl.*, 439(10):3264–3278, 2013.

[39] W. Ding, L. Qi, and Y. Wei. Fast Hankel tensor-vector product and its application to exponential data fitting. *Numer. Linear Algebra Appl.*, 22(5):814–832, 2015.

[40] W. Ding, L. Qi, and Y. Wei. Inheritance properties and sum-of-squares decomposition of Hankel tensors: Theory and algorithms. *BIT Numer. Math.*, 2016. doi: 10.1007/s10543-016-0622-0.

[41] W. Ding and Y. Wei. Generalized tensor eigenvalue problems. *SIAM J. Matrix Anal. Appl.*, 36(3):1073–1099, 2015.

[42] W. Ding and Y. Wei. Solving multi-linear systems with $\mathcal{M}$-tensors. *J. Sci. Comput.*, 2016. doi: 10.1007/s10915-015-0156-7.

[43] W. Ding and Y. Wei. Tensor logarithmic norms and the stability of a class of nonlinear dynamical systems. *Numer. Linear Algebra Appl.*, 2016. to appear.

[44] J. Doyle. Analysis of feedback systems with structured uncertainties. *Proc. IEE-D*, 129(6):242–250, 1982.

[45] L. Eldén and B. Savas. A Newton-Grassmann method for computing the best multilinear rank-(r_1, r_2, r_3) approximation of a tensor. *SIAM J. Matrix Anal. Appl.*, 31(2):248–271, 2009.

[46] L. Elsner. Inverse iteration for calculating the spectral radius of a non-negative irreducible matrix. *Linear Algebra and Appl.*, 15(3):235–242, 1976.

[47] D. Fasino. Spectral properties of Hankel matrices and numerical solutions of finite moment problems. In *Proceedings of the International Conference on Orthogonality, Moment Problems and Continued Fractions (Delft, 1994)*, volume 65, pages 145–155, 1995.

[48] S. Friedland, S. Gaubert, and L. Han. Perron-Frobenius theorem for nonnegative multilinear forms and extensions. *Linear Algebra Appl.*, 438(2):738–749, 2013.

[49] G. H. Golub and C. F. Van Loan. *Matrix Computations*. Johns Hopkins Studies in the Mathematical Sciences. Johns Hopkins University Press, Baltimore, MD, fourth edition, 2013.

[50] L. Grasedyck, D. Kressner, and C. Tobler. A literature survey of low-rank tensor approximation techniques. *GAMM-Mitteilungen*, 36(1):53–78, 2013.

[51] J. He and T.-Z. Huang. Inequalities for M-tensors. *J. Inequal. Appl.*, 2014(1):114, 2014.

[52] D. J. Higham and N. J. Higham. Structured backward error and condition of generalized eigenvalue problems. *SIAM J. Matrix Anal. Appl.*, 20(2):493–512, 1999.

[53] N. J. Higham. *Accuracy and Stability of Numerical Algorithms*. Society for Industrial and Applied Mathematics (SIAM), Philadelphia, PA, second edition, 2002.

[54] C. J. Hillar and L.-H. Lim. Most tensor problems are NP-hard. *J. ACM*, 60(6):Art. 45, 39, 2013.

[55] R. A. Horn and C. R. Johnson. *Topics in Matrix Analysis*. Cambridge University Press, Cambridge, 1994. Corrected reprint of the 1991 original.

[56] S. Hu, Z. Huang, and L. Qi. Strictly nonnegative tensors and nonnegative tensor partition. *Sci. China Math.*, 57(1):181–195, 2014.

[57] S. Hu, Z.-H. Huang, C. Ling, and L. Qi. On determinants and eigenvalue theory of tensors. *J. Symbolic Comput.*, 50:508–531, 2013.

[58] S. Hu and L. Qi. Algebraic connectivity of an even uniform hypergraph. *J. Comb. Optim.*, 24(4):564–579, 2012.

[59] S. Hu and L. Qi. The Laplacian of a uniform hypergraph. *J. Comb. Optim.*, 29(2):331–366, 2015.

[60] S. Hu and L. Qi. A necessary and sufficient condition for existence of a positive Perron vector. *SIAM J. Matrix Anal. Appl.*, 2016. to appear.

[61] S. Hu, L. Qi, and J.-Y. Shao. Cored hypergraphs, power hypergraphs and their Laplacian H-eigenvalues. *Linear Algebra Appl.*, 439(10):2980–2998, 2013.

[62] S. Hu, L. Qi, and J. Xie. The largest Laplacian and signless Laplacian H-eigenvalues of a uniform hypergraph. *Linear Algebra Appl.*, 469:1–27, 2015.

[63] Z. Jia, W.-W. Lin, and C.-S. Liu. A positivity preserving inexact Noda iteration for computing the smallest eigenpair of a large irreducible M-matrix. *Numer. Math.*, 130:645–679, 2015.

[64] X.-Q. Jin. *Developments and Applications of Block Toeplitz Iterative Solvers*, volume 2 of *Combinatorics and Computer Science*. Kluwer Academic Publishers Group, Dordrecht; Science Press Beijing, Beijing, 2002.

[65] T. Kailath and R. H. Roy III. ESPRIT–estimation of signal parameters via rotational invariance techniques. *Optical Engineering*, 29(4):296–313, 1990.

[66] T. G. Kolda and B. W. Bader. Tensor decompositions and applications. *SIAM Rev.*, 51(3):455–500, 2009.

[67] T. G. Kolda and J. R. Mayo. Shifted power method for computing tensor eigenpairs. *SIAM J. Matrix Anal. Appl.*, 32(4):1095–1124, 2011.

[68] T. G. Kolda and J. R. Mayo. An adaptive shifted power method for computing generalized tensor eigenpairs. *SIAM J. Matrix Anal. Appl.*, 35(4):1563–1581, 2014.

[69] D. Kressner and C. Tobler. Krylov subspace methods for linear systems with tensor product structure. *SIAM J. Matrix Anal. Appl.*, 31(4):1688–1714, 2010.

[70] C. Li, F. Wang, J. Zhao, Y. Zhu, and Y. Li. Criterions for the positive definiteness of real supersymmetric tensors. *J. Comput. Appl. Math.*, 255:1–14, 2014.

[71] G. Li, L. Qi, and Q. Wang. Are there sixth order three dimensional PNS Hankel tensors? *arXiv preprint arXiv:1411.2368*, 2014.

[72] G. Li, L. Qi, and Y. Xu. SOS-Hankel tensors: Theory and application. *arXiv preprint arXiv:1410.6989*, 2014.

[73] G. Li, L. Qi, and G. Yu. The Z-eigenvalues of a symmetric tensor and its application to spectral hypergraph theory. *Numer. Linear Algebra Appl.*, 20(6):1001–1029, 2013.

[74] W. Li and M. K. Ng. On the limiting probability distribution of a transition probability tensor. *Linear Multilinear Algebra*, 62(3):362–385, 2014.

[75] Z. Li, H.-T. Huang, Y. Wei, and A. H.-D. Cheng. *Effective Condition Number for Numerical Partial Differential Equations*. Alpha Science International Ltd, UK, 2014 and Science Press, Beijing, second edition, 2015.

[76] L.-H. Lim. Singular values and eigenvalues of tensors: A variational approach. In *IEEE CAMSAP 2005: First International Workshop on Computational Advances in Multi-Sensor Adaptive Processing*, pages 129–132, 2005.

[77] F. T. Luk and S. Qiao. A fast singular value algorithm for Hankel matrices. In *Fast algorithms for structured matrices: theory and applications (South Hadley, MA, 2001)*, volume 323 of *Contemp. Math.*, pages 169–177. Amer. Math. Soc., Providence, RI, 2003.

[78] Z. Luo, L. Qi, and N. Xiu. The sparsest solutions to Z-tensor complementarity problems. *Optim. Lett.*, 2016. doi: 10.1007/s11590-016-1013-9.

[79] Z. Luo, L. Qi, and Y. Ye. Linear operators and positive semidefiniteness of symmetric tensor spaces. *Sci. China Math.*, 58(1):197–212, 2015.

[80] J.-G. Luque and J.-Y. Thibon. Hankel hyperdeterminants and Selberg integrals. *J. Phys. A*, 36(19):5267–5292, 2003.

[81] I. L. MacDonald and W. Zucchini. *Hidden Markov and Other Models for Discrete-valued Time series*, volume 70 of *Monographs on Statistics and Applied Probability*. Chapman & Hall, London, 1997.

[82] Y. Matsuno. Exact solutions for the nonlinear Klein-Gordon and Liouville equations in four-dimensional Euclidean space. *J. Math. Phys.*, 28(10):2317–2322, 1987.

[83] M. Ng, L. Qi, and G. Zhou. Finding the largest eigenvalue of a nonnegative tensor. *SIAM J. Matrix Anal. Appl.*, 31(3):1090–1099, 2009.

[84] M. K. Ng. *Iterative Methods for Toeplitz Systems*. Numerical Mathematics and Scientific Computation. Oxford University Press, New York, 2004.

[85] G. Ni, L. Qi, and M. Bai. Geometric measure of entanglement and U-eigenvalues of tensors. *SIAM J. Matrix Anal. Appl.*, 35(1):73–87, 2014.

[86] Q. Ni and L. Qi. A quadratically convergent algorithm for finding the largest eigenvalue of a nonnegative homogeneous polynomial map. *J. Glob. Optim.*, 61:627–641, 2015.

[87] J. Nocedal and S. J. Wright. *Numerical Optimization*. Springer Series in Operations Research and Financial Engineering. Springer, New York, second edition, 2006.

[88] T. Noda. Note on the computation of the maximal eigenvalue of a nonnegative irreducible matrix. *Numer. Math.*, 17:382–386, 1971.

[89] V. Olshevsky. *Structured Matrices in Mathematics, Computer Science, and Engineering. II.* Contemporary Mathematics, vol. 281. American Mathematical Society, Providence, RI, 2001.

[90] V. Olshevsky and A. Shokrollahi. A superfast algorithm for confluent rational tangential interpolation problem via matrix-vector multiplication for confluent Cauchy-like matrices. In *Structured matrices in mathematics, computer science, and engineering, I (Boulder, CO, 1999)*, volume 280 of *Contemp. Math.*, pages 31–45. Amer. Math. Soc., Providence, RI, 2001.

[91] I. V. Oseledets. Tensor-train decomposition. *SIAM J. Sci. Comput.*, 33(5):2295–2317, 2011.

[92] I. V. Oseledets and E. E. Tyrtyshnikov. Breaking the curse of dimensionality, or how to use SVD in many dimensions. *SIAM J. Sci. Comput.*, 31(5):3744–3759, 2009.

[93] J. M. Papy, L. De Lathauwer, and S. Van Huffel. Exponential data fitting using multilinear algebra: the single-channel and multi-channel case. *Numer. Linear Algebra Appl.*, 12(8):809–826, 2005.

[94] J.-M. Papy, L. De Lathauwer, and S. Van Huffel. Exponential data fitting using multilinear algebra: The decimative case. *Journal of Chemometrics*, 23(7-8):341–351, 2009.

[95] V. Pereyra and G. Scherer. *Exponential Data Fitting and its Applications.* Bentham Science Publishers, 2010.

[96] D. Potts and M. Tasche. Parameter estimation for multivariate exponential sums. *Electron. Trans. Numer. Anal.*, 40:204–224, 2013.

[97] D. Potts and M. Tasche. Parameter estimation for nonincreasing exponential sums by Prony-like methods. *Linear Algebra Appl.*, 439(4):1024–1039, 2013.

[98] L. Qi. Eigenvalues of a real supersymmetric tensor. *J. Symbolic Comput.*, 40(6):1302–1324, 2005.

[99] L. Qi. Eigenvalues and invariants of tensors. *J. Math. Anal. Appl.*, 325(2):1363–1377, 2007.

[100] L. Qi. Symmetric nonnegative tensors and copositive tensors. *Linear Algebra Appl.*, 439(1):228–238, 2013.

[101] L. Qi. H^+-eigenvalues of Laplacian and signless Laplacian tensors. *Commun. Math. Sci.*, 12(6):1045–1064, 2014.

[102] L. Qi. Hankel tensors: associated Hankel matrices and Vandermonde decomposition. *Commun. Math. Sci.*, 13(1):113–125, 2015.

[103] L. Qi, Y. Wang, and E. X. Wu. D-eigenvalues of diffusion kurtosis tensors. *J. Comput. Appl. Math.*, 221(1):150–157, 2008.

[104] M. Rajesh Kannan, N. Shaked-Monderer, and A. Berman. Some properties of strong $\mathcal{H}$-tensors and general $\mathcal{H}$-tensors. *Linear Algebra Appl.*, 476:42–55, 2015.

[105] W. C. Rheinboldt. *Methods for Solving Systems of Nonlinear Equations*, volume 70 of *CBMS-NSF Regional Conference Series in Applied Mathematics*. Society for Industrial and Applied Mathematics (SIAM), Philadelphia, PA, second edition, 1998.

[106] S. Rouquette and M. Najim. Estimation of frequencies and damping factors by two-dimensional ESPRIT type methods. *IEEE Transactions on Signal Processing*, 49(1):237–245, 2001.

[107] R. Roy and T. Kailath. ESPRIT-estimation of signal parameters via rotational invariance techniques. *IEEE Transactions on Acoustics, Speech and Signal Processing*, 37(7):984–995, 1989.

[108] S. M. Rump. Perron-Frobenius theory for complex matrices. *Linear Algebra Appl.*, 363:251–273, 2003. Special issue on nonnegative matrices, M-matrices and their generalizations (Oberwolfach, 2000).

[109] X. Shi and Y. Wei. A sharp version of Bauer-Fike's theorem. *J. Comput. Appl. Math.*, 236(13):3218–3227, 2012.

[110] J. A. Shohat and J. D. Tamarkin. *The Problem of Moments*. American Mathematical Society Mathematical surveys, vol. I. American Mathematical Society, New York, 1943.

[111] R. S. Smith. Frequency domain subspace identification using nuclear norm minimization and Hankel matrix realizations. *IEEE Transactions on Automatic Control*, 59(11):2886–2896, 2014.

[112] Y. Song and L. Qi. Infinite and finite dimensional Hilbert tensors. *Linear Algebra Appl.*, 451:1–14, 2014.

[113] G. W. Stewart and J. G. Sun. *Matrix Perturbation Theory*. Computer Science and Scientific Computing. Academic Press, Inc., Boston, MA, 1990.

[114] W. J. Stewart. *Probability, Markov Chains, Queues, and Simulation*. Princeton University Press, Princeton, NJ, 2009.

[115] J.-G. Sun. *Matrix Perturbation Theory*. Science Press, Beijing, second edition, 2001. (in Chinese).

[116] L. Sun, S. Ji, and J. Ye. Hypergraph spectral learning for multi-label classification. In *Proceedings of the 14th ACM SIGKDD international conference on Knowledge discovery and data mining*, pages 668–676. ACM, 2008.

[117] C. Tobler. *Low-rank tensor methods for linear systems and eigenvalue problems*. PhD thesis, Eidgenössische Technische Hochschule ETH Zürich, Nr. 20320, 2012.

[118] S. Trickett, L. Burroughs, and A. Milton. Interpolation using Hankel tensor completion. In *SEG Technical Program Expanded Abstracts 2013*, pages 3634–3638. 2013.

[119] E. E. Tyrtyshnikov. How bad are Hankel matrices? *Numer. Math.*, 67(2):261–269, 1994.

[120] C. Van Loan. *Computational Frameworks for the Fast Fourier Transform*, volume 10 of *Frontiers in Applied Mathematics*. Society for Industrial and Applied Mathematics (SIAM), Philadelphia, PA, 1992.

[121] R. S. Varga. *Geršgorin and His Circles*, volume 36 of *Springer Series in Computational Mathematics*. Springer-Verlag, Berlin, 2004.

[122] X. Wang and Y. Wei. $\mathcal{H}$-tensors and nonsingular $\mathcal{H}$-tensors. *Front. Math. China*, 11:557–575, 2016.

[123] Y. Wang, J.-W. Chen, and Z. Liu. Comments on "Estimation of frequencies and damping factors by two-dimensional ESPRIT type methods". *IEEE Transactions on Signal Processing*, 53(8):3348–3349, 2005.

[124] Y. Wang, G. Zhou, and L. Caccetta. Nonsingular H-tensor and its criteria. *J. Ind. Manag. Optim.*, 12(4):1173–1186, 2016.

[125] C. Xu. Hankel tensors, Vandermonde tensors and their positivities. *Linear Algebra Appl.*, 491:56–72, 2016.

[126] W. Xu and S. Qiao. A fast symmetric SVD algorithm for square Hankel matrices. *Linear Algebra Appl.*, 428(2-3):550–563, 2008.

[127] S. Yan, D. Xu, B. Zhang, H.-J. Zhang, Q. Yang, and S. Lin. Graph embedding and extensions: a general framework for dimensionality reduction. *IEEE Trans. Pattern Anal. Mach. Intell.*, 29(1):40–51, 2007.

[128] Q. Yang and Y. Yang. Further results for Perron-Frobenius theorem for nonnegative tensors II. *SIAM J. Matrix Anal. Appl.*, 32(4):1236–1250, 2011.

[129] Y. Yang and Q. Yang. Further results for Perron-Frobenius theorem for nonnegative tensors. *SIAM J. Matrix Anal. Appl.*, 31(5):2517–2530, 2010.

[130] Y. Yang and Q. Yang. *A Study on Eigenvalues of Higher-Order Tensors and Related Polynomial Optimization Problems*. Science Press, Beijing, 2015.

[131] L. Zhang and L. Qi. Linear convergence of an algorithm for computing the largest eigenvalue of a nonnegative tensor. *Numer. Linear Algebra Appl.*, 19(5):830–841, 2012.

[132] L. Zhang, L. Qi, and Y. Xu. Linear convergence of the LZI algorithm for weakly positive tensors. *J. Comput. Math.*, 30(1):24–33, 2012.

[133] L. Zhang, L. Qi, and G. Zhou. M-tensors and some applications. *SIAM J. Matrix Anal. Appl.*, 35(2):437–452, 2014.

[134] T. Zhang. Existence of real eigenvalues of real tensors. *Nonlinear Anal.*, 74(8):2862–2868, 2011.

[135] G. Zhou, L. Qi, and S.-Y. Wu. Efficient algorithms for computing the largest eigenvalue of a nonnegative tensor. *Front. Math. China*, 8(1):155–168, 2013.

[136] D. Zwillinger. *Handbook of Differential Equations*. Academic Press, Inc., Boston, MA, third edition, 1997.

Subject Index

A

Absolute backward error, 19
Absolute M-equations, 106–107
Anti-circulant tensors, 45–51
  block tensors, 49–50
  diagonalization, 46–47
  singular values, 47–49
Augmented Vandermonde
    decomposition, 68–75

B

Backward error analysis, 19–20
Banded M-Equation, 107–108
Bauer-Fike theorem, 33–34, 35
BHHB tensor. *See* Block Hankel tensor
    with Hankel blocks (BHHB
    tensor)
Block compressed generating vector, 49
Block generating vector, 49
Block Hankel tensor with Hankel blocks
    (BHHB tensor), 43, 44, 45,
    49, 52

C

CANDECOMP/PARAFAC (CP)
    decomposition, 8
Cauchy matrices, 61
Collatz-Wielandt formula, 27–29, 34
Componentwise backward error, 20
Conuent Vandermonde matrix, 69
Convolution formula, Hankel tensor,
    61–63
Crawford number, 21–22

D

Decomposition
  augmented Vandermonde, 68–75
  CANDECOMP/PARAFAC (CP), 8

  Hankel tensor, 65–66
  SOS, 65–66
  tensor, 8–10
  Tucker-type, 8, 9
Diagonal equation, 99
Diagonalizable tensor pairs, 15–16
Dirichlet's boundary condition, 98, 122
Discrete-time Markov chain, 5

E

Eigenvalue problem
  generalized matrix, 11
  tensor, 10
Exponential data fitting, 40
  multidimensional case, 42–45
  one-dimensional case, 40–42

F

Fast Fourier transformations (FFT), 40
Fast Hankel tensor-vector product,
    50–53
Fourier matrix, 46, 62

G

Gauss-Seidel method, 109
Generalized matrix eigenvalue
    problem, 11
Generalized tensor eigenvalue
    problem, 10
  backward error analysis, 19–20
  definitions, 13–14
  diagonalizable tensor pairs, 15–16
  Gershgorin circle theorem, 16–19
  number of eigenvalues, 14–15
  properties, 14–20
  spectral radius, 15
  unified framework, 11–13
Gershgorin circle theorem, 16–19, 34, 36

H

Hankel matrices, 61
Hankel tensor, 39–40, 51
    convolution formula, 61–63
    decomposition of, 65–66
    inheritance properties, 59–77
    lower-order implies higher-order,
        63–65
    strong, 66–68
    Vandermonde decomposition of,
        67–75
Hankel tensor-vector products, 40, 42
    average running time of, 54
    numerical examples, 53–57
Hankel total least squares (HTLS)
        method, 40
H-eigenvalue, 21, 22
Higher order orthogonal iterations
        (HOOI), 42, 55
Higher-order singular value
        decomposition (HOSVD), 9,
        39, 41
Higher-order tensor, 7
Hilbert tensor, Vandermonde
        decomposition for, 73
Homogeneous tensor eigenvalue
        problem, 10
Homogenous polynomial dynamical
        system, 13
HOOI. *See* Higher order orthogonal
        iterations (HOOI)
HOSVD. *See* Higher-order singular
        value decomposition
        (HOSVD)
HTLS method. *See* Hankel total least
        squares (HTLS) method
Hypermatrix. *See* Tensor

I

Inheritance properties, Hankel tensor,
        59–77
Inverse iteration, 118–122

L

Laplacian tensor, 12, 13
Least squares (LS), 40

M

Markov chain, discrete-time, 5
Matrix eigenproblem, 108
M-equation, 102–104
    absolute, 106–107
    backward errors of triangular,
        115–116
    banded, 107–108
    classical iterations, 109
    condition numbers, 116–118
    iterative methods of, 108–114
    Newton method for symmetric,
        111–112
    nonhomogeneous left-hand side,
        105–106
    nonpositive right-hand side,
        104–105
    numerical tests, 112–114
    perturbation analysis of, 114–118
M-matrix, 81–83
Monotonicity, 90
    definitions, 90
    example, 93
    nontrivial monotone Z-tensor, 93
    properties, 90–92
M-tensor, 81–83
    extension of, 93–95
    general, 89–90
    spectral properties of, 83–84
Multilinear equations, 97

N

Newton method, for symmetric
        M-equations, 111–112
Noda iteration, 118
Nonnegative tensor, 81–82
Normwise backward error, 19, 20

O

Optimization problem, 11

P

Papy's method, 42
Perron-Frobenius theorem, 27, 81, 82, 84
Polynomial dynamical system,
        homogenous, 13

R
Rayleigh-Ritz theorem, 34
Real tensor pairs, 20
    Crawford number, 21–22
    symmetric-definite tensor pairs,
        22–26
Relative backward error, 19
R-regular tensor pair, 21

S
Semi-positivity
    definitions, 84–85
    Z-tensor, 85–87
Semi-symmetric tensor, 18
Sign-complex spectral radius, 26
    Bauer-Fike theorem, 33–34
    Collatz-Wielandt formula, 27–29
    componentwise distance to
        singularity, 31–32
    definitions, 26–27
    single tensors properties, 29–31
Single tensors properties, 29–31
Singular value decomposition (SVD), 8
SOS
    decomposition of strong Hankel
        tensor, 65–66
    tensor, 59, 61
Spectral radius, 15
    sign-complex, 26–34
Strong Hankel tensor, SOS
    decomposition of, 65–66
SVD. *See* Singular value decomposition
    (SVD)
Symmetric-definite tensor pairs, 22–26
Symmetric *M*-equation, Newton method
    for, 111–112

T
Takagi factorization, 65, 66
Taylor expansion tool, 5
Tensor, 3–6. *See also specific types of*
    tensor
    anti-circulant, 45–51
    BHHB, 43, 44, 45, 49, 52
    block, 49–50
    decompositions, 8–10

    eigenvalue problems, 10
    Frobenius norms of slices of core, 57
    higher-order, 7
    Hilbert, 73
    Laplacian, 12, 13
    nonnegative, 81–82
    operations, 6–8
    semi-symmetric, 18
    single, 29–31
    SOS, 59, 61
    summation, 95–96
Tensor-matrix multiplications, 7
Tensor pair
    diagonalizable, 15–16
    R-regular, 21
    symmetric-definite, 22–26
Tensor train (TT) format, 10
TLS. *See* Total least squares (TLS)
Toeplitz matrices, 45, 61
Total least squares (TLS), 40
Triangular equations, 99–102
Triangular *M*-equation, backward errors
    of, 115–116
Tucker-type decomposition, 8, 9
2D ESPRIT method, 44
2D fast Fourier transformation
    (FFT2), 52

U
US-eigenvalues, 12

V
Vandermonde decomposition, 41, 43
    Hankel tensor, 67–75
    Hilbert tensor, 73
Vandermonde matrix, 41, 44, 61, 62
    conuent, 69

Y
Yule-Walker equation, 68

Z
Z-matrix, 82
Z-tensor
    nontrivial monotone, 93
    semi-positivity, 85–87

Printed in the United States
By Bookmasters